MW00782901

Graph Vision

Graph Vision

Digital Architecture's Skeletons

Theodora Vardouli

The MIT Press
Cambridge, Massachusetts
London, England

For Aliki and Dimitri
with whom this book has grown

Contents

1 Graphs 1
2 Images 13
  Peril 13
  Forms 15
  Textbooks 25
  Frameworks 48
  Isomorphisms 74
3 Tools 79
  Ends 79
  Data 80
  Layouts 87
  Patterns 106
  Generativities 128
4 Infrastructures 133
  Control 133
  Change 140
  Choice 150
  Possibility 155
  Combinatorics 180
5 Skeletons 185
  Ghosts 185
  Flesh 187
  Boxes 188
  Closets 190

Acknowledgments 193
Notes 197
Index 221

1 Graphs

Among Blanche Descartes's endeavors in mathematical invention and literary whimsy was the 1969 poem "The Expanding Unicurse."[1] A peculiar interlude in the proceedings of a mathematical conference at the University of Michigan, the poem chronicled the trajectory of a mathematical *bête noire* that Blanche had helped nurture to fame. The verse began with the inception of a distinctive mathematical object by Swiss polymath Leonhard Euler as he riddled through the infamous crossings over the river Pregel in Königsberg, mashing churches and cathedrals, passages and homes, Kant's erstwhile promenades, into adimensional abstractions.[2] Four capital letters for masses of land and seven lowercase letters for the bridges that connected them birthed the first *graph*: a set of points representing discrete objects and a set of lines representing these objects' relations. An exciting invention, the graph was nonetheless met with scorn and disrespect by early twentieth-century mathematicians for its filiation with frivolous problems—a condition to which, some say, Blanche owed her existence.[3]

Appearances, you see, deceive. If graphs afforded a new vision that lifted the disguise of Königsberg's architectural topography to reveal a truthful hidden structure, Blanche was all about disguise. A witty combinatorics of the initials of a foursome of male mathematicians at the University of Cambridge, Blanche was a playful ruse for work on graphs when they were still derided—and, as the pseudonym reveals, feminized—as recreational mathematics.[4] And yet by the 1960s, as Blanche's poetic verse declared, graphs had spread their tendrils into

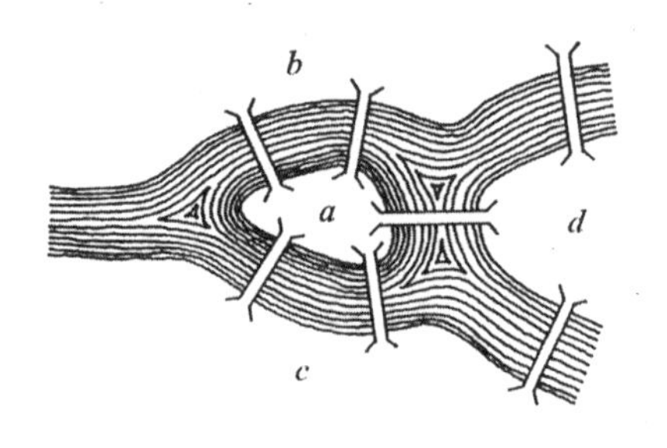

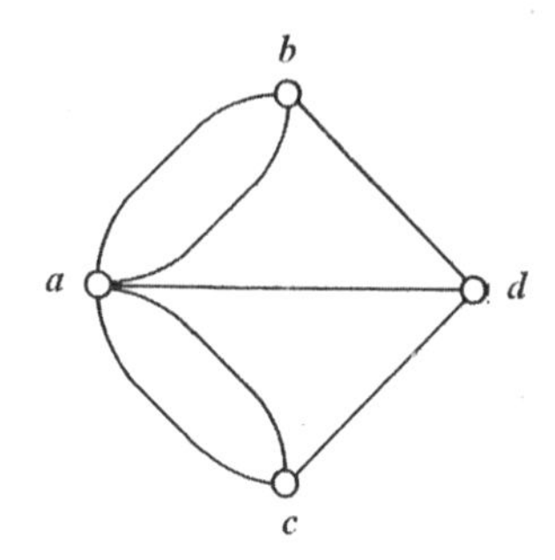

Figure 1.1
The Königsberg bridge problem, graphs' celebrated origin story. *Source*: Lionel March and Philip Steadman, *The Geometry of Environment: An Introduction to Spatial Organization in Design* (Cambridge, MA: MIT Press, 1974), 242. Copyright Philip Steadman and the Estate of Lionel March.

high-powered research universities and percolated through numerous fields outside mathematics. In architecture, graphs had a brilliant and transformative career. They produced new images, tools, and infrastructures that captured the architectural imagination and left thundering echoes in their wake: echoes that continue to reverberate, but perhaps now require more attentive ears to be discerned. This book is an architectural history of graphs when they were at their loudest.

This history is also a prehistory. "Digital architecture"—the medley of architectural software programs, computational design processes, and the built outcomes of their use—has skeletons hidden in its closet. Or rather, under its surface. That is, underneath complex architectural geometries, and behind the screens of proliferating software programs for designing, managing, and constructing them, lurk skeletal abstractions that consist of discrete entities and their relationships. Architects who work digitally are continually, although often implicitly, confronted with the manipulation of the abstract structures through which digital electronic computers parse architectural processes. Skeletal structures of points and lines or webs and networks are also familiar promotional imagery for digital architecture conferences and publications. And more recently, manifestos for "discrete" architecture have been casting the combination of virtual and physical components in calculable structures as a new aesthetic and political agenda for architecture.[5]

One might be tempted to interpret digital architecture's reliance on skeletal structures as an effect of how digital electronic computers work. Digital computers mostly operate with elemental units and structured relational descriptions. This book rejects the proposition that the dovetailing of architecture and structural abstraction—on a cultural, discursive, and operational level—is an effect of digital computers' technical functioning. Instead, it contributes a multifaceted story of architects' turn toward the structural underpinnings of their work as the very condition that made it possible to consider architectural design as a form of computation. By making structures operative, graphs configured basal, and persistent, agendas of representation, instrumentality, and political agency tied to digital computers.

Uncanny doppelgängers, graphs and skeletons work to abstract and construct architectural worlds through points and lines, distinct entities and relations. But if skeletons are stealthily concealed, graphs were prominently displayed. They were drawn, sketched, printed, and etched in the pages of controversial, yet impactful, publications that promised a computational future for architecture: one that relied on precise, transparent methods for processing information, producing form, and anticipating its performances. This book is an iconology of this prolific visual production. It disentangles graphs' fluctuating aesthetic, epistemic, and political meanings as they moved across sites of architectural research. Instead of considering graphs as instruments for achieving the calculative necessities of nascent computational approaches to design and architecture, the book casts them as catalysts of a transformative encounter between architecture, mathematics, and digital computing.

But how to fix, and why fixate on, a mathematical entity? And how to cast the graph as an icon without inadvertently turning it into a fetish; without exaggerating its power or flattening the diverse contexts of its manifestation? A degree of what one might call "methodological fetishism"—focusing on things and their trajectories—in part distinguishes this book from other histories of architecture, mathematics, and computation that have granted institutions, human actors, concepts, or agendas primary analytical significance.[6] Instead, the book centers on the graph as a protean but distinct mathematical entity and considers its immutabilities, mutations, and mobilities. More than simply a vehicle for structuring a historical narrative, the graph here figures as a historical agent that shaped the aims architects enlisted it to support.[7] In other words, this is not a story about how architects applied mathematics to serve various, and often dubious, purposes. Disciplinary designations and stable assignments of intentionality are useful but fall short in periods of epistemic tumult and invention. The book takes the graph instead as an object of work and affection and acknowledges its status as "embodying concepts": as an "epistemic" (and in our case also an aesthetic, discursive, operative) "thing."[8]

Fixating on graphs allows us to bring together eccentric mathematicians, research proselytizers, architectural anxieties, academic cults, legendary feuds, dramatic failures, and informational optimisms into a larger project of renegotiating architecture's representations, instrumentalities, and politics. But fixating on the graph is easier than fixing it. For, as we will see, the graph's architectural mobilities were precisely fueled by the very conditions that architects enlisted graphs to obliterate—ambiguity, seductiveness, and disguise. The story of graph vision is therefore a story of profound ambivalence toward visual appearance as a site of architectural knowledge, operation, and value production.

§

Picture this: a constellation of infinitesimal points scattered on a surface and threaded together with lines; straight, sinuous, spiraling, and meandering. Now imagine the surface of this webbed drawing deforming, expanding, stretching, twisting, and contracting. Every deformation makes a new image: dimensions shift, symmetries collapse, figures dissolve. But these visual dissimilarities matter little. Behind this visual polyphony lies a stable object. The graph you pictured remains the same despite drastic changes in its geometric appearance. Such is the condition of graph vision: hidden sameness behind sense-perceptible difference; visual variety undergirded by structural invariants.

This book is concerned with the infolding of this condition in architectural representation to produce a set of images that purported to reveal what lay *under*, *behind*, or *beyond* visual appearance and geometric apprehension. It sheds light on the rhetorical and operative valences of this mode of representation by paying attention to the material and aesthetic attributes of graphs as things that moved among architectural knowledge settings.[9] In doing so, it unveils historical connections among geographically and institutionally dispersed research groups and organizations, but also brings them together as parts of the same story: a story of the construction of a new vision of, and for, architecture.

To talk of vision is to tread treacherous terrain. Among its many definitions, vision denotes the physiological action of seeing, the act of seeing by means other than ordinary sight, and the ability to conceive what might be attempted or achieved. It is at once appearance, apparition, and envisioning.[10] My discussion of graphs connects to a contemporary concern with "visualization" as a form of knowledge production, of argument and rhetoric, and of action.[11] Indeed, graphs often held representational or generative relationships with information of various kinds. But graphs were not always, or at least not exclusively, about data. As we will see, graphs were always, also, about shape—about architecture's representations as those had been coded by Euclidean geometry.

Corrective and oppositional, the relationship between graphs and geometric shapes opened the path toward a new visual modality in architecture: one that cast visual appearance (and its corollaries of subjectivity, arbitrariness, seductiveness, and deception) as a manifestation of a nonvisual structure that could, however, be rendered visible. The architectural vision modeled on and facilitated by graphs did not dispel

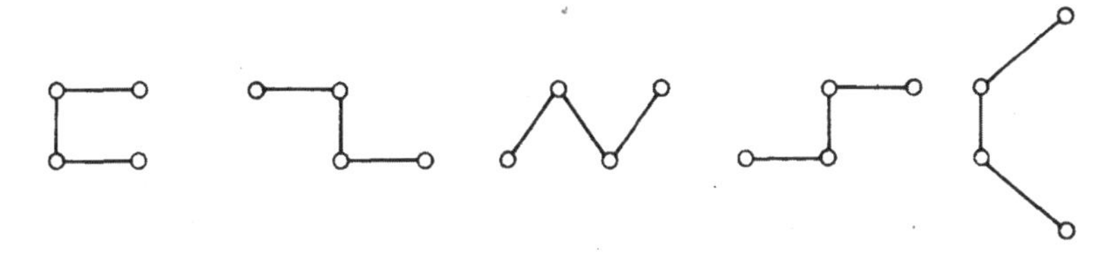
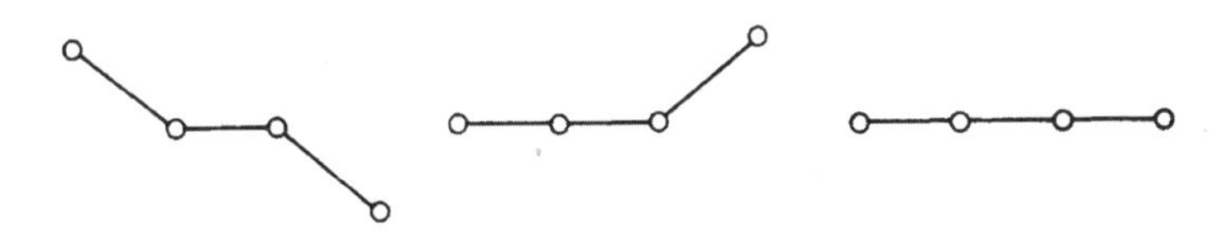

Figure 1.2
Eight ways of drawing a graph of four points. The eight drawings are all the same graph. *Source*: Frank Harary, "Aesthetic Tree Patterns in Graph Theory," *Leonardo* 4, no. 3 (1971): 228.

the visible world, but rather cast mathematically calculable abstract structures both as the generators of sense-perceptible appearances and as their limit.

Attention to this condition's entanglements with discourse, technics, and fledgling subjectivities motivates the book's iconological approach. The book assembles and interprets an iconography of graphs. My use of iconography and iconology is unapologetically loose: readers should not expect a rigorous exposition of graph images' multiple strata, as Erwin Panofsky would want it.[12] Instead, I use the terms to signal a commitment to graphs' status as images that carried meanings in architectural communities of knowledge and practice. These meanings were not intrinsic to the images, but contingent on their settings of production and use. In that sense, the images' content—the topologies of the graphs and the ways these were visually rendered—factor in the interpretation as much as their intellectual, material, and institutional contexts.

Graph vision, the shorthand for the architectural visuality, as Hal Foster might have called it, that emerged from the contested relationship between graphs and shapes, also speaks to the shifting condition of architectural representation with the advent of computer graphics and the proliferation of digital electronic computers.[13] Recently, scholars have marshaled the metaphor of "vision" to encapsulate the technical, material, and labor conditions that birthed abstractions and codifications of geometry in early computer graphics systems, but also to signal these processes' productiveness in ushering in new *visions* of design—its methods and subjects.[14] Others have challenged the interpretation of computer graphics as computationally enhanced visual images, arguing for computer graphics as producing their own invisibility—one that does not work through revealing, but instead through hiding.[15] Concerned with an ambivalence between revealing and hiding, surface and structure, the concretely phenomenal and symbolically abstract, the book moves from the iconography of graphs to the concealment of the skeleton.

Alongside shaping grounds for critically contemplating architecture's digital theories and techniques, the book contributes to histories of architectural representation and its twentieth-century reconfigurations. The graph osculates with the "diagram," a notoriously elusive keyword in contemporary architectural theory pointing to a part-visual, part-conceptual entity that gathers forces and influences motivating architectural form. Graphs are a kind of diagram and were often referred to as such. But although this is a book on graphs, this is not a book on the architectural diagram. I strategically set aside this paramount disciplinary discourse to consider graphs through historical excavation rather than theoretical rumination.

Much of the discussion around the "diagram" that animated architectural theory around the 1990s arose precisely from a desire to distinguish the term from its midcentury precedents, including graphs. Robert Somol, for instance, drawing from Peter Eisenman, famously dismissed such precedents as "inadequately diagrammatic" because of their use as "essentialist tools" aspiring to formal and operational truths.[16] The objection was not to the operative, conceptual, or representational qualities of these earlier diagrammatic attempts, but rather to the ways they were mobilized discursively.

Retheorized as diagrams, graphs continued to enjoy unabashed affection from heralds of the so-called digital turn in architecture.[17] Indulgent, if not imprecise, definitions of these diagrams as Deleuzian "abstract machines"—nimble relationships of forces—supported many of their formal and informational excursions.[18] In a sense, diagrams of data-driven projects are to graphs what poststructuralism is to structuralism: theoretical realignments performed upon technical continua. Using graphs instead of diagrams as a historical category acknowledges their mathematical meanings and provenance: their existence and signification at the nexus of architectural and mathematical cultures. Graphs enter the story through transactions and transitions of architectural and mathematical modernism, in their assimilation of one of the postwar period's most enthralling intellectual projects: the redescription of disciplinary knowledge in structural terms.

§

The end of the 1950s found architects who adhered to the tenets of interwar modern architecture grappling with what seemed like an inheritance of questionable architectural conventions rather than actionable methods.[19] In the aftermath of historic events such as the closing of the Bauhaus in 1933 and the demise of the Congrès Internationaux d'Architecture Moderne (CIAM) in 1959, proponents of modern architecture's values of science and rationality and advocates of its socially oriented, problem-solving ethos sought out new processes for accomplishing these persisting agendas.[20] A new generation of architects saw the teachings of their predecessors as perpetuating baseless dogmas. Acolytes of modern architecture who populated academic institutions disseminated a theory that had lost its "original validity as an approach to the creation of form" and had "dwindled into the decorative by its retention of absolute concepts belonging to outworn systems of thought."[21] Despite their tone, such critiques testified to modern architectural theory's staying power, as opposed to its demise: the aim was revision, not rejection.[22]

This revisionist and powerful "second modernism," as scholars have dubbed the transmutations of modern architectural ideas and ideals before the postmodernism of the 1970s, was closely linked to the knowledge institution of the research university.[23] In the United States and the United Kingdom, the research university emerged as a privileged site of knowledge production after the Second World War, because of its relative seclusion from the exigencies of practice and its alignment with principles and ideals of basic research.[24] The knowledge setting of the research university served as a site for architects to robustly engage with mathematics. This continued earlier twentieth-century traditions in which the demands of mass production had motivated scrutiny of proportional systems, dimension standardization, and modular coordination, granting mathematics meanings in architecture that ranged from the quasi-metaphysical to the utterly pragmatic.[25] Similarly, commitments to economy and efficiency had promoted perusal of statistical averages and circulation distances. The new surge in architectural theory during the 1960s did not deploy more of the same mathematics. Instead, architectural theorists and researchers turned to a new kind of mathematics that was then entering public and intellectual debates in full force.

Continuing to undergo a modernist transformation initiated in the late nineteenth century, mathematics came to symbolize a distinctive disciplinary attitude characterized by a formal, abstract, and inward-looking attitude toward its subject matter.[26] This was the result of fervent efforts to rebuild the mathematical edifice in a rigorous and coherent way, or, as Nicolas Bourbaki put it, to "[tear] down the old sections with their labyrinths of alleys, and [project] towards the periphery new avenues, more direct, broader and more commodious."[27] Nicolas Bourbaki was the moniker of a secret mathematical collective that played a paramount role in promoting mathematical modernism. Borrowed from a Russian general, the pseudonym Bourbaki figures as a militaristic, male counterpart to our Blanche.[28] If Blanche wrote waggish prose to elevate overlooked mathematical processes and objects, Bourbaki aimed to tear down mathematics and rebuild it from the ground up. Its members championed a vision of the discipline as a unified, expanding structure built up from combinations of mathematical entities. The entities or their properties were not important. Instead, what mattered were their relations and combinations.[29]

This meant shifting the focus of mathematics from the study of specific mathematical objects and their properties to studying relations between mathematical structures that connected them. This continued an abstracting project initiated by German mathematician David Hilbert, who in his 1899 *Foundations of Geometry* famously espoused Euclid's axiomatic method as a valid model of constructing and organizing mathematical truths but reworked it to correct what he perceived as a weakness: many of Euclid's theorems were based on spatial intuitions about points, lines, and planes rather than on abstract reason. In a move frequently viewed as a watershed moment for mathematical modernism, Hilbert dissociated the fundamental objects of geometry from any spatial and visual properties and endeavored to produce a geometry whose statements would hold true even if points, lines, and planes were replaced by "tables, chairs, and beer mugs."[30]

Bourbaki's approach similarly signaled a shift from what the objects of a field's study are or what they do to the structures that constitute them or of which they are part. This move toward "structure" also had implications outside mathematics, rendering Bourbaki both catalyst and reference point for broader intellectual transformations.[31] The physical sciences but also anthropology, sociology, archaeology, literature, and other fields that previously had relied on empirical knowledge emulated Bourbaki's move toward structural abstraction. "Structure" became a stand-in for disciplinary renewal, modernization, and rigorization, and a keyword in the intellectual tide of structuralism that swept the so-called "soft" sciences after the Second World War.[32] In the late 1960s comparative and historical sociologist Garry Runciman was questioning whether structuralism had managed to amount to a "distinctive doctrine or method" of analysis, while philosopher Michel Foucault spoke of it as "not a new method" but as "the awakened and troubled consciousness of modern thought."[33] Despite intellectual and discursive polyphony, structural approaches to various knowledge domains often relied on mathematics such as set theory, matrix theory, topology, and graph theory, which broadly represented structures and whose study was boosted by mathematics' modernist transformation.

In architecture, the "mathematics of quality," as French anthropologist Claude Lévi-Strauss referred to the modern mathematics of relations and structures, fostered new optimism for a truly modern architectural discipline with well-organized knowledge and a solid mathematical foundation.[34] Graphs supported the tenet of studying relations instead of properties of objects, a premise rife with cultural connotations of modernization and unification. But as proponents of graph theory incessantly advertised, graphs' geometric appearance offered advantages of accessibility and congeniality. Graphs made structures and relations *visible* and *workable*, without that visibility compromising their abstract, intellectual content. Especially for architecture, traditionally entrenched in geometry and cultures of seeing, the Janus-faced graphs fostered hopes of reconciling the subjective and objective parts of the discipline, or to use two fond yet misty keywords, "intuition" and "rationality."

§

The points and lines of graphs could do remarkable things: by stripping dissimilar things to their bare-bone skeletons, they could visually reveal them to be deep down the same; by assigning weights to the lines and labels to the points, they could help predict how different configurations of rooms would perform in relation to stated criteria, and by counting all the possible ways a fixed set of points (and the abstract entities they represented) could be combined with lines (relationships), graphs could generate and exhaustively enumerate all possible configurations of a particular thing composed of these entities. In the 1960s and 1970s, architects invested these operations with visions of reforming a wavering modernism that appeared in need of immediate intellectual and methodological update.

This reformative impulse seeped through architecture with its promise of salvaging modern architecture's intellectual project of rigor and reason by steering away from appearances and stylistic preferences. Architects turned toward mathematics in search of a solid disciplinary ground. Gradually, architecture became infiltrated with talk of systems, structures, trees, networks, and patterns along with a wide gamut of graph varieties to represent and implement them. The book's chapters tell a story of three uses of graphs: graphs as *images*, graphs as *tools*, and graphs as *infrastructures*. These three chapters are linked to the ontology of the graph in strong ways: each is a call back to three mathematical and computational properties—*isomorphism*, *generativity*, and *combinatorics*—inextricably tied to the graph's perceived epistemic, instrumental, and political affordances.

The chapter "Images" grapples with the graph as a new kind of architectural drawing—in its materiality, expression, and dissemination—that was nonetheless argued *not* to be a drawing. Graphs were portrayed as free from the perils and contingencies of traditional orthographic drawings that were fastened to the relentless specificities of a drawing surface, of geometric attributes and measurements, and of their author's graphic style. By unpacking the term *isomorphism*, an architectural keyword borrowed from mathematics to denote "equality of form" between two dissimilar things, the chapter considers claims around graphs' *unlikeness* to architectural drawings and the functions of their visual *likeness* to architectural drawings and other contemporaneous representations of architectural space. The chapter "Tools" shows that graphs served not only as new kinds of drawings, but also as tools for *generating* new architectural drawings. It engages two distinct uses of graphs in the context of efforts to algorithmically automate architectural design (with or without the use of computers): first graphs as tools for matching structures of activity and architectural space, and second graphs as tools for structuring design processes themselves. "Infrastructures" then probes the rhetorical repositioning of graphs from technorational tools to an infrastructure for open-ended choice in the context of participatory design and do-it-yourself architecture theories and programs. There, material and mathematical skeletons bootstrapped each other, co-constructing what I refer to as an "infrastructure model" of design: the provision of a fixed, invariant structure on which to superimpose personal preferences and meanings.

If once celebrated on screen and on paper, graphs have now retreated to the background of various digital tools for architectural design—a pervasive but invisible calculative underbelly. The book ends at this moment of concealment, of naturalization and effacing. And yet graph vision persists. Behind the scrumptiously continuous geometries once associated with "digital architecture," or their more recent replacement with assemblies boisterously showcasing their components, lie discrete representations of entities and relationships. It seems that we have come full circle. Lauded for their ability to purify architectural representation of its visual aspects, graphs proliferated as a new mode of seeing, describing, and designing architecture. Ironically, it was precisely *drawings* of graphs that captured architects' attention. Aligning with the logics of digital electronic computers, graphs also brought architecture into the computer as a potentially automatable process. Graphs appeared on the screen and then hid behind it, from where they continue to puppeteer images of architectural geometries that refresh on our monitors. The book ends with graphs' haunting of architecture's presents, with skeletons and their receptacles.

2 Images

Peril

It was time to pass the baton. Le Corbusier was 74, one year younger than Mies, Gropius was 78, and Wright had died two years ago. But this wasn't about age. The start of the 1960s found the sermons of interwar modern architecture echoing loudly in the ears of architects, but with nagging doubts about their relevance in the face of galloping technological change. Criticism was emerging about modern architecture's adherence to "powers from the mains and the reduction of machines to human scale," to repeat Reyner Banham's pithy dismissal, and its proponents' inability to keep up with a technological culture that seemed to have gone on without them.[1] As suspected since the 1930s, architectural modernism seemed to have disintegrated into a style of austere geometry and industrial materials that its followers mimetically emulated.[2] An update seemed to be in order. The epigones of the Bauhaus or the recently dismantled CIAM had to take stock and move forward.

Such was the impetus for a series of presentations and discussions held from 20 March to 1 May 1961 at Columbia University. Titled "The Four Great Makers and the Next Phase in Architecture," the six-week event self-advertised as at once a celebration, a call, and a plea.[3] The program featured Le Corbusier, Walter Gropius, Mies van der Rohe, and (in spirit) the late Frank Lloyd Wright—the "great makers" who, in Dean Charles Colbert's words, "furnished the bedrock upon which all contemporary architecture rests."[4] Throughout the program's eight "cycles," celebratory convocations and gala dinners as homage to the "great makers" alternated with anxious questions such as "conformity, chaos, or continuity?," "cul-de-sac or open end?," and "obsolete or viable?" from their successors—international architects, educators, and writers invited to reexamine and reformulate architecture's central issues.[5]

Among the participants of the fifth cycle on "The House for the Modern Family" was Serge Chermayeff, a Russian émigré architect who had come to Yale University after directing the Institute of Design in Chicago—an outgrowth of László Moholy-Nagy's "New Bauhaus"—and serving as professor at Harvard's Graduate School of Design during the chairmanship of Bauhaus founder Walter Gropius.[6] With his address, initially announced under the title "The Socially-Disciplined Housing of Walter Gropius," Chermayeff chastised his fellow panelists for "very eloquently propound[ing]" "aesthetic principles."[7] "Architects," he continued, "still voice piously their belief in the[se] aesthetic principles . . . and then parody these principles miserably in acts utterly contemptible."[8]

Chermayeff portrayed his contemporaries as "hiding their artistic, ostrich-like necks in the sand where decision and power lie elsewhere" and urged shifting attention to modern architecture's "programmatic," as he called it, inaugural agenda—the one that it carried before being reduced to an aesthetic style.[9] "The kind of fashionable millinery which distinguishes our present architectural era," he declared, "surely calls again for the formulation of principles as serious as those which originally moved the men honored at these exercises."[10] Pointing to CIAM's founding agenda "to see to the resolution of architectural problems," Chermayeff implored: "Let Us Not Make Shapes: Let Us Solve Problems."[11]

It's a telling slogan. This is not only because of its functionalist rhetoric and moral subscription to the idea of architects as agents of social betterment. It is about the denouncing of "shapes" as a treacherous force. Was Chermayeff championing architectural iconoclasm? Or did the negation imply an alternative, a shift from drawing shapes and thinking with shapes to another mode of representation and reasoning? One can only speculate. Yet by 1961 Chermayeff had been exposed to inklings of a new method for doing architecture that displaced geometric shapes as the privileged realm of architectural invention. This was through the work of his doctoral student Christopher Alexander, a Cambridge University-trained mathematician and architect who had moved to Harvard for PhD studies in 1958.

In spring 1961, around the time of the "Four Great Makers" events, Alexander was presenting at the Building Research Institute a preliminary method for designing in a systematic way. Alexander had begun developing this method during a two-year collaboration with Chermayeff at the Harvard-MIT Joint Center for Urban Studies.[12] The method revealed what Alexander claimed was a "non-arbitrary" structure for organizing design requirements. This structure mandated the sequence by which an architect could address these requirements to produce architectural form. At the Building Research Institute conference, Alexander first publicly announced a mathematical device that would allow designers to "*picture* [emphasis mine]" this structure: the "topological 1-complex" or, more simply, the linear graph.[13] The graph represented the complex relationships between the requirements, the ways that they affected each other. It also made it possible, through calculations performed on these relationships, to order the graph in a hierarchical structure, a "tree," that prescribed the order in which a designer ought to tackle them. The designer started by designing for simple groups of requirements at the bottom and moved "up" the tree, combining these partial drawings as the tree indicated to produce the full design. The tree also indicated the organization of a library of building-related information that should be made available to the designer for a specific design brief.

The graph was, as Alexander said, a *picture*, an image. But it was not a shape. It was what lay under architectural appearance and governed its geometric articulation. Its mathematical properties evaded metric specificities. We may then read Chermayeff's plea as something more than polemical iconoclasm. It was a declaration of ambivalence toward images and seeing shared by various interlocutors who collectively ushered a new visual culture in architecture founded upon a paradoxical negation of the visual. For all the devaluing of the surface in favor of unveiling abstract hidden structures, we will see that it was photographs in elementary school textbooks and drawings in experimental architectural magazines, in their unapologetic visualness and materiality, that boosted architecture's rising skeletal representations.

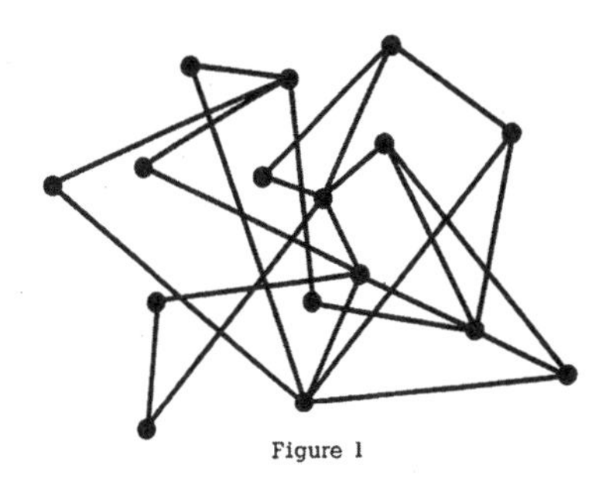

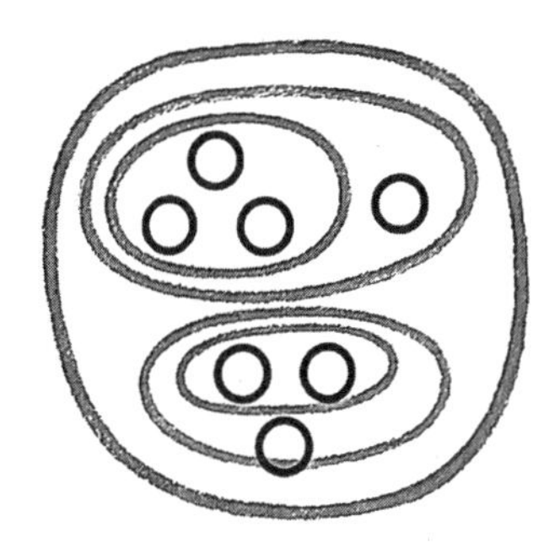

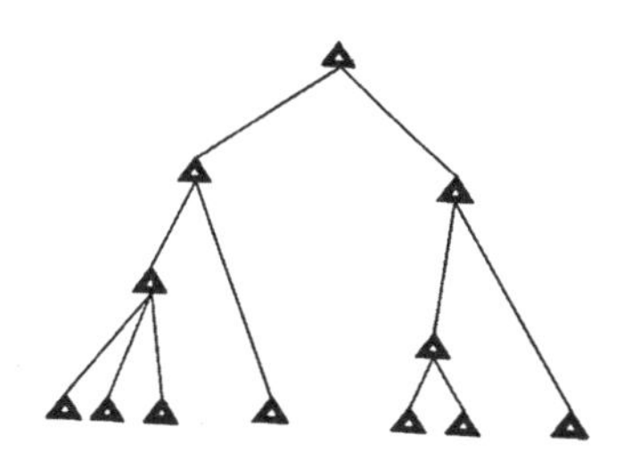

Figure 2.1
Diagrams from a new process for "picturing" the structure of a design problem. *Source*: Christopher Alexander, "Information and an Organized Process of Design," in *New Building Research Spring 1961* (Washington, DC: National Academy of Sciences—National Research Council, 1962), 118. Courtesy of the Christopher Alexander / Center for Environmental Structure Archives.

Forms

The ambivalence toward the visual is perhaps best captured in a tongue-in-cheek manifesto published in *Architectural Design* in 1971: "Draughtsmanship is a drug. Without vision we may indeed perish, but the vision we need is to see, in a hard intellectual light, things as they are; not to have hallucinations."[14]

The members of the Land Use Built Form Studies (LUBFS) Centre at the University of Cambridge who coauthored the manifesto sermonized a "structural revolution" that would radically overhaul the mathematical foundation of architectural representation and, consequently, reasoning. Their declaration, though, was no plea for architectural typhlosis. Instead, it was about rendering visible and operational what would otherwise be intangible and conceptual. It was about a new mode of architectural seeing delivered by a new roster of architectural images. Beacons of intellectual clarity were to be found in "modern mathematics," revealing, in the direct sense of *picturing*, "structural patterns of objectivity" behind "surface subjectivity."[15] Immediately after the diatribe against the "drug" of "draughtsmanship," a declaration of distrust and caution against the treachery of architectural appearances, came an oft-raised dilemma about architects' responsibility: "Individual expression, or social service? Personal prejudice, or communal enquiry?"[16]

The manifesto, opening a special issue of *Architectural Design* on the work of the LUBFS Centre since its foundation by Sir Leslie Martin in 1967, was self-conscious rhetoric. LUBFS Centre director Lionel March and members Peter Dickens and Marcial Echenique intentionally wrote the introduction in the language of a 1930s architectural manifesto.[17] Abundant in military metaphor, the manifesto recalled the heroic period of architectural modernism, a modernism undermined by the treachery of appearance. For it was the attachment to appearance that had caused the modern project to fail, or so Leslie Martin had argued a few months before founding the LUBFS Centre in a presentation to the Royal Institute of British Architects (RIBA), published as "Architect's Approach to Architecture."[18]

In his talk, Martin assessed that it was not the principles of modern architecture that had led to its demise, but architects' mistaking of geometric systems produced by these principles as design doctrines.[19] Modern architecture, he tellingly suggested, mandated that buildings ought to be "thought out rather than drawn."[20] Doctrinaire adherence to geometric systems countered what Martin saw as an essential element of the modern attitude: the constant reassessment of architectural problems through rationality and technical innovation. To reclaim this spirit, Martin laid out an ambitious program of rethinking relationships between "patterns of use" and "pattern of form" through analysis, experiment, and testing.[21]

Interest in the relationship between form and use had grown out of a series of studies on the redesign of Whitehall, London's old government district, a large and controversial project commissioned to Martin in 1964.[22] Martin had recruited a small team of researchers to build a mathematical model that would investigate the effects of built form on the efficiency of land use and explore more possibilities than the voguish building forms of the tower or the slab.[23] As geometric and mathematical explanations of the relationship between built forms and land use became clearer, Martin began to envision applying the same

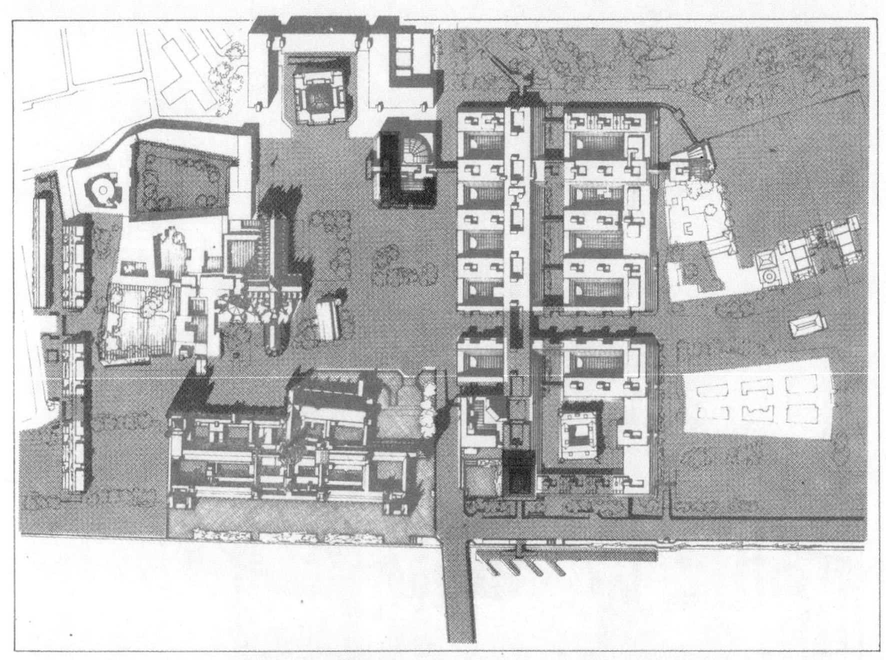

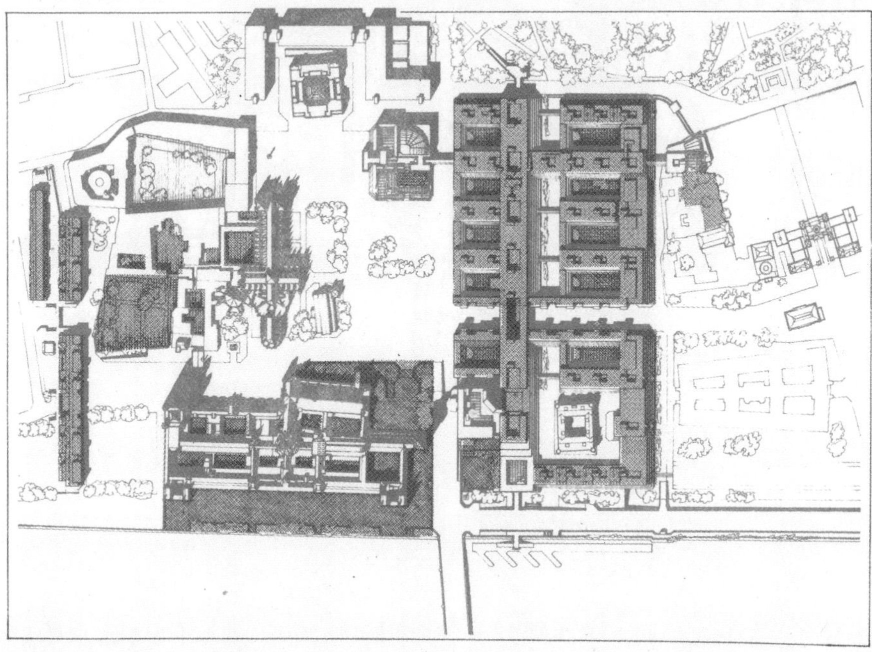

Figure 2.2
Proposal report for the redevelopment of Whitehall, London's government district. *Source*: Lionel March, *Whitehall: A Plan for a National and Government Centre Report* (London, 1964), reprographic copy, 30 × 42 cm, ARCH287273. Courtesy of the Lionel March fonds, Canadian Centre for Architecture, Gift of Candida March, © Estate of Lionel March.

Figure 2.3
Whitehall study. Area calculations for different forms. *Source*: Lionel March, *Whitehall: A Plan for a National and Government Centre Report* (London, 1964), reprographic copy, 30 × 42 cm, ARCH287273. Courtesy of the Lionel March fonds, Canadian Centre for Architecture, Gift of Candida March, © Estate of Lionel March.

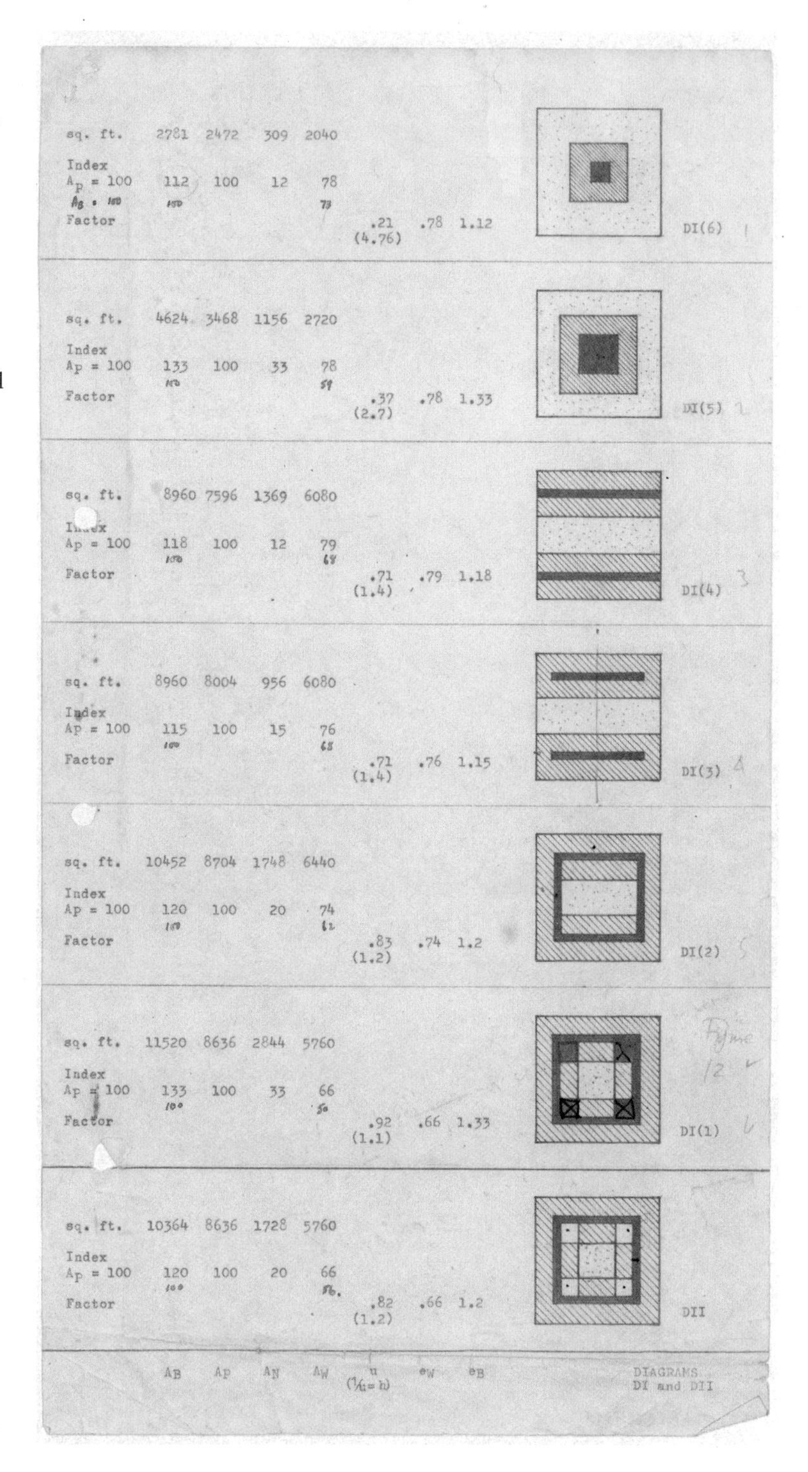

	A_B	A_P	A_N	A_W	u ($1/u = h$)	e_W	e_B	DIAGRAMS DI and DII
sq. ft.	2781	2472	309	2040				DI(6)
Index A_P = 100	112	100	12	78				
Factor					.21 (4.76)	.78	1.12	
sq. ft.	4624	3468	1156	2720				DI(5)
Index Ap = 100	133	100	33	78				
Factor					.37 (2.7)	.78	1.33	
sq. ft.	8960	7596	1369	6080				DI(4)
Index Ap = 100	118	100	12	79				
Factor					.71 (1.4)	.79	1.18	
sq. ft.	8960	8004	956	6080				DI(3)
Index Ap = 100	115	100	15	76				
Factor					.71 (1.4)	.76	1.15	
sq. ft.	10452	8704	1748	6440				DI(2)
Index Ap = 100	120	100	20	74				
Factor					.83 (1.2)	.74	1.2	
sq. ft.	11520	8636	2844	5760				DI(1)
Index Ap = 100	133	100	33	66				
Factor					.92 (1.1)	.66	1.33	
sq. ft.	10364	8636	1728	5760				DII
Index A_P = 100	120	100	20	66				
Factor					.82 (1.2)	.66	1.2	

principles at a larger scale. These research pursuits went hand in hand with a broader agenda that he was vigorously promoting during his chairmanship at Cambridge: to academically establish architecture by developing "theory"—a body of principles organizing the discipline's knowledge base and informing practice.[24] Motivated by such aspirations, Martin founded the LUBFS Centre in 1967 as the first research center of the Department of Architecture at Cambridge University.[25]

Martin found a keen supporter of this quest for theory, rigor, and mathematical explanation in his former student and Whitehall study collaborator Lionel March, whom he appointed director of the Centre in 1969. March was one of three students with a background in mathematics to join Martin's first class, along with Christopher Alexander and William (Bill) Newman, who later had a prolific career in computer science.[26] A mathematical prodigy praised by computing herald Alan Turing for a project on individual numbers he had done independently while still in school, March studied mathematics at Cambridge under Dennis Babbage, but found the topics stale.[27] Instead, he spent most of his first year at Trinity College designing sets, costumes, and graphics for plays mounted by the Amateur Dramatic Club in the mid-1950s, such as John Millington Synge's *Deidre of the Sorrows* and Sophocles' *Philoctetes*. Set and costume design eventually led March to the Department of Architecture, to which he was mostly attracted because of Martin's connections with Bauhaus émigrés and a burgeoning cultural scene, exemplified for instance by his coeditorship, with Naum Gabo and Ben Nicholson, of the single-issue magazine *Circle: International Survey of Constructivist Art*.[28]

March's alignments with avant-garde art went together with a commitment to clearheaded and scientifically driven inquiry. He was a vocal advocate of research and rational analysis, which he argued would provide an expanding bedrock of knowledge in support of architectural education and "establish the foundations of an academic discipline."[29] For instance, in two opinion articles, one published in the *RIBA Journal* in 1972 and one in the *Cambridge Review* in 1973, March variously celebrated academia's fortuitous distance from the tumult of professional decisions and compromises and the "unique opportunity" for a "long, cool view" that it offered.[30] He positioned the LUBFS Centre among the "many skirmishes" that had, for a decade or so, been establishing a "beachhead for . . . the 'mathematization' of the human sciences."[31]

By the time the two articles appeared, the LUBFS Centre had published numerous working papers and articles, whose topics ranged from representing spatial structure in cities, to allocating activities and circulation in offices and universities, to generating architectural floor plans based on dimensional and functional requirements.[32] These projects had either a footing or an application in real-world scenarios, a condition stemming in part from the Centre's sponsorship by external philanthropic, national, or governmental organizations.[33] Underlying the variety of scales and foci in the Centre's projects was a desire to expand architecture's reach beyond the scale of the building and implant it at the epicenter of the "environmental sciences"—a broad term that encompassed all fields studying aspects of the human-made and natural environment in its material, social, anthropological, and other dimensions.[34]

March had been exposed to the contested position and potentials of architecture for catalyzing urban research before arriving at the LUBFS Centre. He held a two-year visiting research appointment at the Harvard-MIT Joint Center for Urban Studies as a Harkness Fellow to study the architectural and urban work of Frank Lloyd Wright—one of March's major inspirations. At the Joint Center March found himself surrounded by interdisciplinary teams of researchers who tackled trans-scalar issues.[35] This gave him a model for research. During his directorship at the LUBFS Centre, March ventured to consolidate architecture's role within similar multiscale and interdisciplinary endeavors. But this pursuit also required a set of working devices. For the LUBFS Center, these devices were mathematical models.

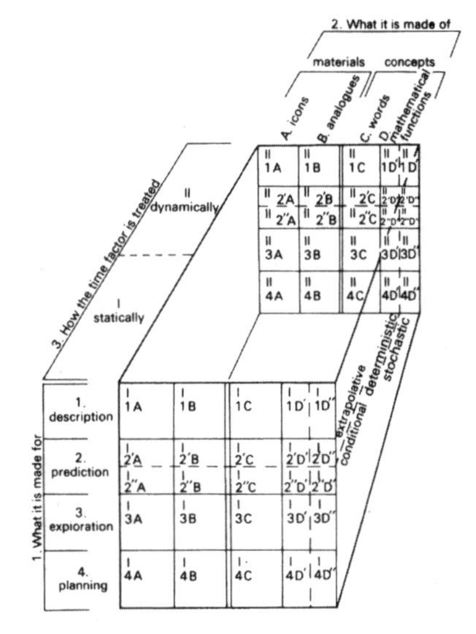

Figure 2.4
Classification system for models with three axes (intention, materials, treatment of time). *Source*: Marcial Echenique, "Models: A Discussion," in *Urban Space and Structures*, ed. Leslie Martin and Lionel March (New York: Cambridge University Press, 1972), 169.

Part empirical, part theoretical, a model was "a representation of a reality," expressing characteristics of this reality deemed pertinent by the modelmaker with material or conceptual means.[36] A mathematical model, then, was a type of model where the "description of reality is represented by the use of symbols and the relationships expressed in terms of operations."[37] The operative term here was *isomorphism*. Taken from mathematics, the term denotes a mapping between two structures that preserves entities and relationships. To create a mathematical model, LUBFS Centre researchers identified architectural entities and relationships and mapped them into mathematical ones. This one-to-one mapping meant that observations made in the mathematical model could be transferred back to its reference system, to architecture.

The reproduction of physical situations turned mathematical models into useful tools for experimental testing of relationships between "environmental situations" and "kinds of actions," "as full-scale experiments with real buildings" were "plainly limited since mistakes will be costly, and . . . identical circumstances . . . rarely repeated."[38] "Built forms," in that sense, were mathematical models of buildings that helped researchers grapple with the complexity and uniqueness of actual buildings by isolating the aspects of building that were relevant for a given study.[39] Mathematical models would establish correspondences and mappings between architecture modeled as "built forms," large institutional buildings modeled as "urban subsystems," and the city modeled as an "urban system," rendering them as problems

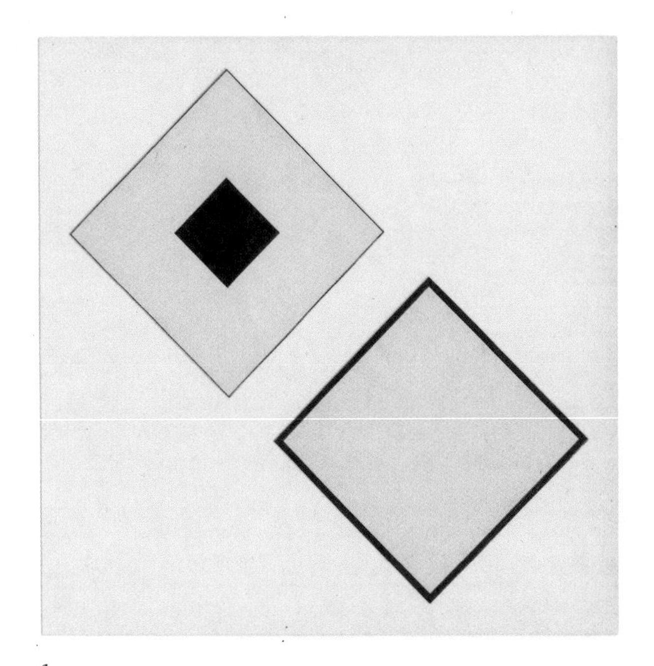

a

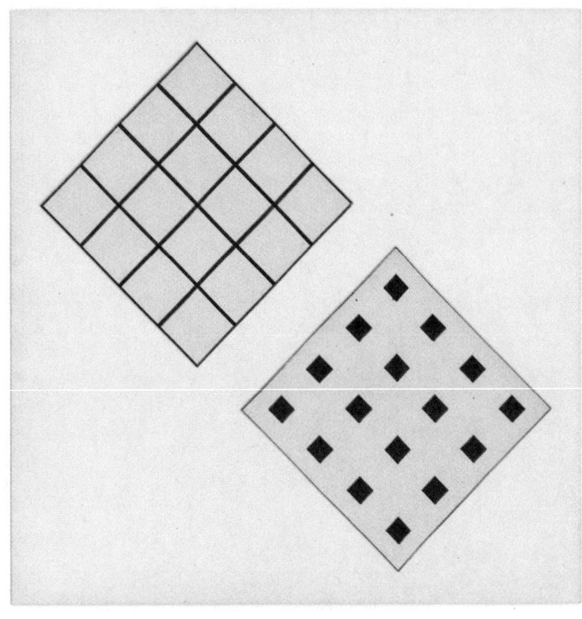

b

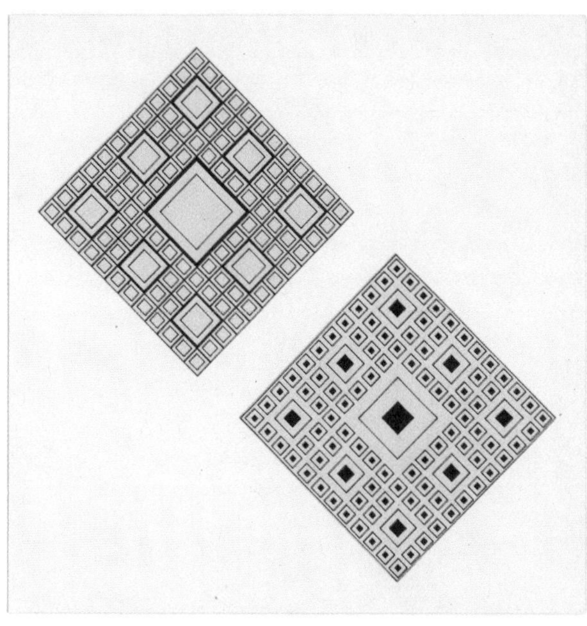

c

Figure 2.5
Patterns of squares in which the ink takes different shapes but covers the same area; example *a* became the logo of the Land Use Built Form Studies Centre. Centre for Land Use and Built Form Studies panels, 1967–1973, collage on cardboard, each 50.5 × 50.5 cm. *Source*: Lionel March Fonds, Canadian Centre for Architecture, ARCH287264 (*a*), ARCH287266 (*b*), ARCH287265 (*c*). Courtesy of the Lionel March Fonds, Canadian Centre for Architecture, Gift of Candida March, © Estate of Lionel March.

Interlude 1

Here are floor plans of three buildings designed between 1938 and 1941 by Frank Lloyd Wright but never built. The plan labeled *a* represents a house for a family of $5,000–6,000 income included in *Life* magazine's "Eight Homes for Modern Living," *b* is the floor plan for Hollywood costume designer Ralph Jester, and *c* is the Vigo Sundt House near Madison, Wisconsin.

The plans are redrawn in crisp lines outlining walls and openings, doing away with the elaborate floor motifs that embellished Wright's originals.

Their visual dissimilarity is stark.

The rectangle, the circle, and the triangle suggest three distinct, even incommensurable architectural languages.

But something hides behind the surface.

Place a point in the middle of every separate room. Now add a line linking the points whenever two rooms connect, and you get a graph. It is the same graph for all three buildings, except the Sundt has one more room.

Isomorphism.

Visual difference dissipates under structural sameness.

Wright was designing the same project again and again.

What other striking geometric manifestations could this graph have yielded?

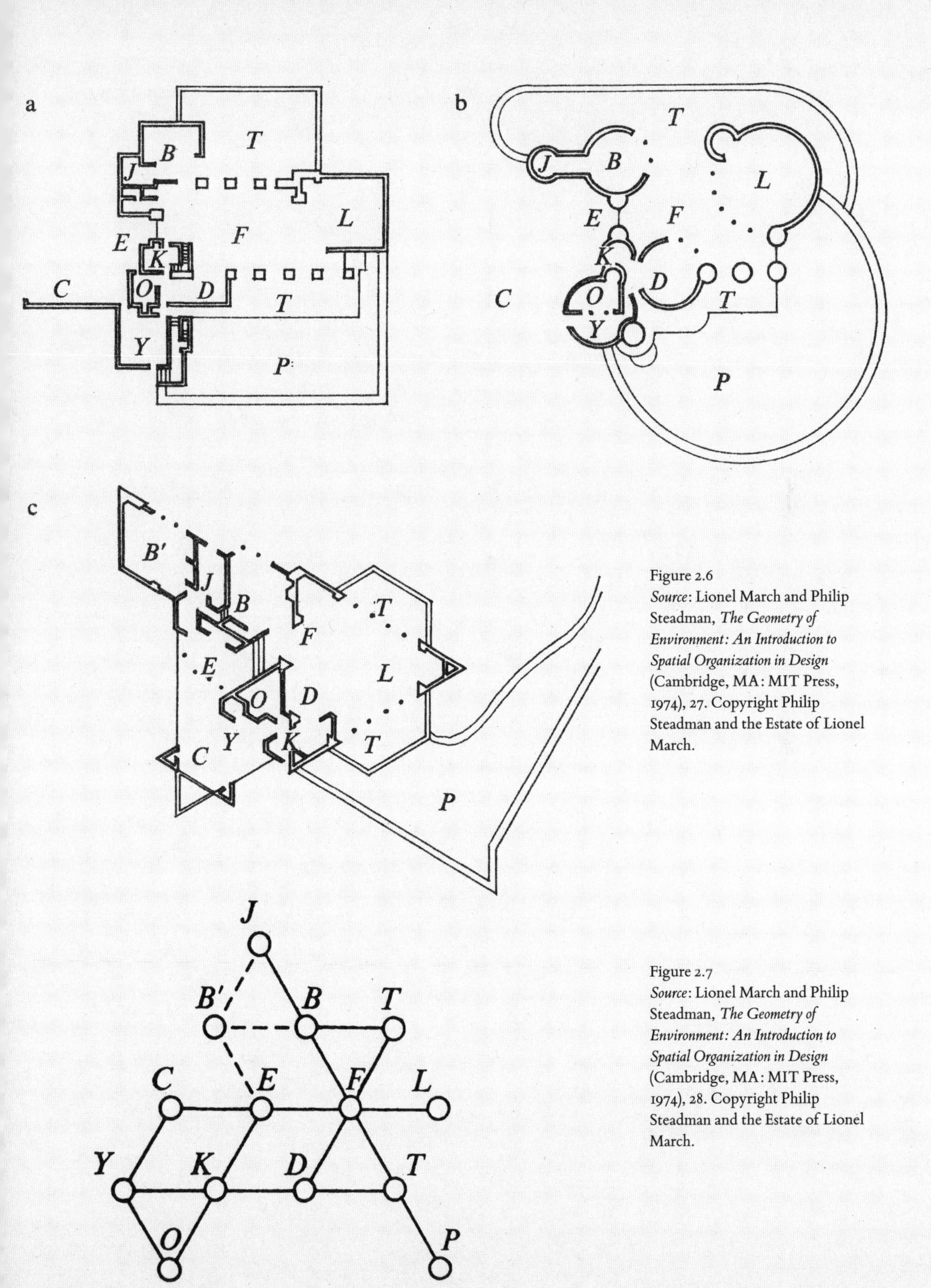

Figure 2.6
Source: Lionel March and Philip Steadman, *The Geometry of Environment: An Introduction to Spatial Organization in Design* (Cambridge, MA: MIT Press, 1974), 27. Copyright Philip Steadman and the Estate of Lionel March.

Figure 2.7
Source: Lionel March and Philip Steadman, *The Geometry of Environment: An Introduction to Spatial Organization in Design* (Cambridge, MA: MIT Press, 1974), 28. Copyright Philip Steadman and the Estate of Lionel March.

of "spatial organization" and placing them all under the purview of architectural research.[40] As Martin argued, different scales and contexts would be integrated as "aspects of the main problem of relationships," and studies starting with a building would extend "right through the whole environmental field."[41]

This integrative impulse permeated the widespread intellectual movements of structuralism and systems theory to which LUBFS Centre researchers pledged allegiance through their pleas for "structural revolutions" and proclamations of the Centre as a "systems laboratory."[42] This was a deliberate alignment with intellectual currents that were infiltrating universities far and wide, and were palpably "in the air" at Cambridge University in fields as diverse as anthropology, economy, geography, sociology, archaeology, and others.[43] Vocal advocates of systems such as influential economist and philosopher Kenneth Boulding—known among other things for his metaphor of planetary harmony and cooperation, "spaceship earth"—promoted them as a remedy to disciplinary balkanization tainting the sciences—a way out of a "desert of mutual unintelligibility" with "walled-in hermits . . . each mumbling to himself words in a private language that only he can understand."[44] Systems, Boulding argued, provided a "skeleton" for science: featureless mathematical entities from which to "hang the flesh and blood of particular disciplines and particular subject matters," thus revealing "similarities in the theoretical constructions of different disciplines."[45]

For March the discovery of isomorphisms among different problems and knowledge domains was a fundamentally aesthetic and creative endeavor. "This is at root an aesthetic activity which the arts and sciences hold in common," he contended.[46] It is tempting to read the term "aesthetic" as pledging faith to the aniconic abstraction of architecture through mathematical symbols and formalisms, as a description of a style of research, or as tasking architectural representation with aestheticizing the vogue of systems.[47] But there is more to it. March's "aesthetic" was tied to a kind of *seeing* appreciative of deep, invisible relationships between things instead of their sense-perceptible surface appearances. It was about cultivating an eye able to discern equalities of form among things that appeared dissimilar at first sight. Forging this seeing produced a new agenda for representation in architecture.

The mathematics enabling architects "to see in a hard intellectual light," to recall the *Architectural Design* manifesto, was not the traditional mathematics of "counting, ordering, comparing, and measuring" but new mathematical varieties: "groups, rings, fields, vector spaces, linear and boolean algebras; topology, graph theory, and varieties of algebraic geometry; linear, non-linear, dynamic and boolean programming."[48] This was no accident or practical necessity. LUBFS Centre work was shaped by cultural valencies of modern mathematics in British education; in turn, it deliberately and consistently moved modern mathematics in the realm of architectural culture.

Textbooks

A peculiar choice, a frontal photograph of the Mishra Yantra printed in sepia on black background, figures on the cover of two reprints of a book written by Lionel March and Philip Steadman in 1971. Initially commissioned by the RIBA and published by the organization's short-lived publishing venture RIBA Publications Limited, *The Geometry of Environment*, as the book was called, was reprinted in 1974 by Methuen Publishing Ltd. and the MIT Press. The reprints were identical to the original, except for the replacement of RIBA Publications' type-only, no-image cover with an altered photograph of the eighteenth-century building-scale astronomical instrument in New Delhi.

Was it lyrical metaphor, earnest didacticism, or a streak of irony that guided this choice? It is tempting to consider the cover as both a statement and a subversion. As an architecture for making calculations, a building that was mathematical and a mathematics that was built, Mishra Yantra crisply embodied the book's argument that architecture and mathematics were inextricably intertwined. But could it also be that the photograph quietly suggested a polemic against an era that the book declared as past, or, perhaps more strongly, that it aspired to end? A line from the book's "Guide to Further Reading" quoting Italian philosopher Benedetto Croce's characterization of aesthetic measures and proportional systems as the "astrology of number" somewhat buttresses this conjecture.[49] When looking at the cover, a device for measuring the heavens striking in its symmetry and seductive curvatures, one ought to see not occult symbolisms projected upon mathematical systems but instead *spatial organization*—a term featured in the book's subtitle to denote an emphasis on relations and structures underlying and organizing geometric form and space.

Even though "geometry" was featured in the book's title, habitual ideas from Euclidean geometry that architects typically used to describe shapes, their appearance, and their metric properties were sparse. Instead, the book presented readers with more unfamiliar subjects such as mappings, transformations, symmetry groups, matrices, modules, planar graphs, and networks. In the book's introduction, coauthors March and Steadman championed the benefits of the architectural reader's exposure to this new kind of geometry that was not about measure but instead about structure and organization:

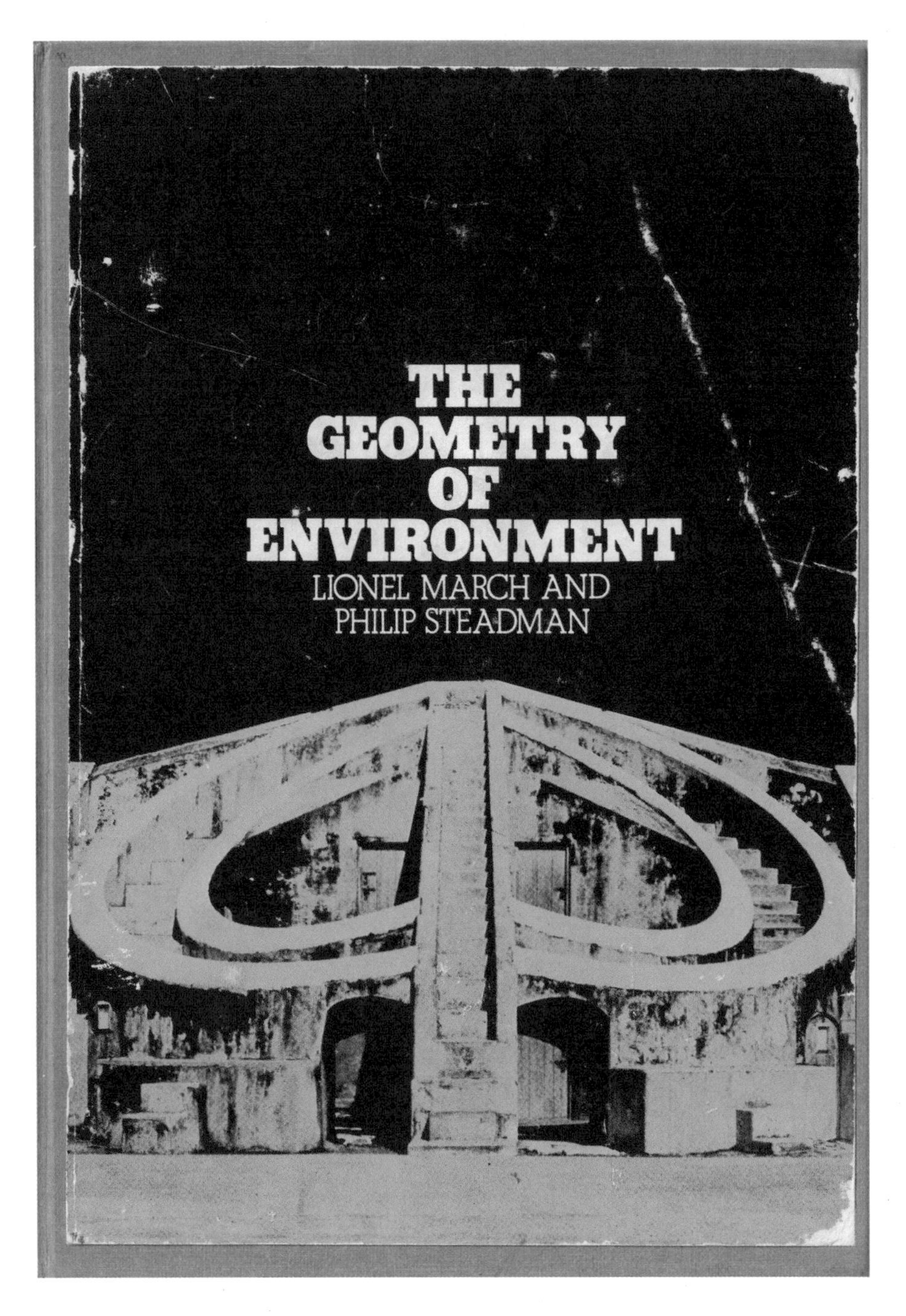

Figure 2.8
The building-scale astronomical instrument Mishra Yantra on the cover of Lionel March and Philip Steadman's *The Geometry of Environment*, originally published by RIBA Publications Ltd. and reprinted by the MIT Press. *Source*: Lionel March and Philip Steadman, *The Geometry of Environment: An Introduction to Spatial Organization in Design* (Cambridge, MA: MIT Press, 1974). Copyright Philip Steadman and the Estate of Lionel March.

> Perhaps the chief difference between the traditional treatment of geometry in architecture and the one presented here, is that, previously, geometry was employed to *measure* [emphasis in the original] properties of space such as area, volume, angle, whereas the new mathematical theories of sets, groups, and graphs—to name but a few—enable us to describe *structural relationships* [emphasis in the original] which cannot be expressed in metrical forms, for example, 'adjacent to', 'in the neighborhood of', 'contained by.'[50]

The Geometry of Environment had a different flavor than the LUBFS Centre working papers and the "Models of Environment" special issue in *Architectural Design*, which appeared in the same year. Glancing at the table of contents, readers may have thought they were reading a mathematical textbook, with each of the book's 14 chapters named after a mathematical concept. The book's first half, written by March, used architectural examples to illustrate new mathematical concepts. The second half, written by Steadman, demonstrated applications of new mathematical ideas to architecture and displayed more prominently material from the Centre's working papers.

The book contained a staggering 319 figures. Most of them were drawn by Catherine Cooke, an eminent scholar of Russian architecture and planning who at the time was working on her dissertation under March's supervision. March and Steadman also produced some of the drawings. The graphic language is familiar and consistent: crisp lines made with a technical pen and bold black fills on thick board. In the book, architectural orthographic drawings are made to look like mathematical diagrams, and mathematical diagrams have something of the rendering of an architectural drawing.

If "Models of Environment" emanated a fervor for architecture's mathematization, *The Geometry of Environment* was written with scholastic restraint. The book conveyed a sense of timeliness, urgency even, but the urgency did not have to do with problems vexing the architectural profession. It was an urgency to contend with remarkable transformations in a surprising site: the school desk. In the decade preceding the book's publication, the subjects covered in *The Geometry of Environment* had come to the foreground of British mathematical education by a roster of new syllabi produced under the so-called "new maths" reforms.[51] The new syllabi integrated twentieth-century developments in mathematics, such as set theory, combinatorics, and matrix algebra—in short, the mathematical techniques that March and Steadman presented in their book.[52]

The book was commissioned to March by the RIBA Library Committee following ongoing conversations about the relationship of mathematics and architecture in the mid-twentieth century. These

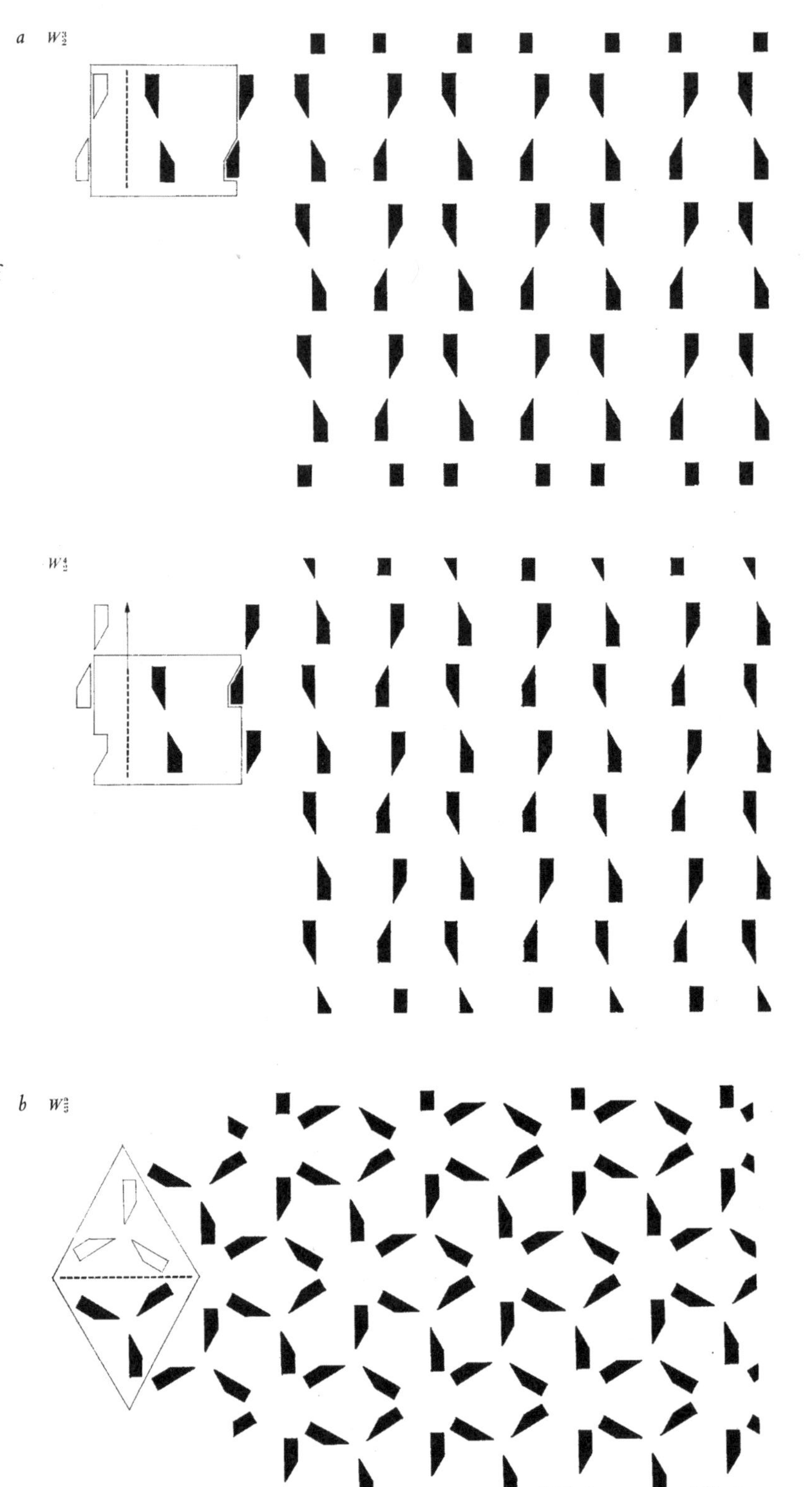

Figure 2.9
Illustrations of symmetry groups. *Source*: Lionel March and Philip Steadman, *The Geometry of Environment: An Introduction to Spatial Organization in Design* (Cambridge, MA: MIT Press, 1974), 69 (*a*), 71 (*b*). Copyright Philip Steadman and the Estate of Lionel March.

conversations were fueled by late 1940s influential architecture historical works, such as those of Colin Rowe and Rudolf Wittkower, which reappraised the architectural thought of the Renaissance by unearthing its mathematical underpinnings.[53] Stimulated by such works, Alison and Peter Smithson, British architects and leaders of the architectural group Team 10, expressed to the RIBA an interest in examining the relationship of mathematics to architecture in the mid-twentieth century.[54] Alongside this query, the Smithsons brought to the RIBA Library Committee's attention a "generational gap" in the mathematics taught to young British students, including their son, which left them perplexed.[55] Because of his mathematical training and connections to the RIBA, March was invited, as he recounted, to "write a book for architects that would illustrate the potential of the 'new maths' in their field."[56]

Introduced by Martin as "a book about the new mathematics and architecture," *The Geometry of Environment* aspired to establish both generational and disciplinary bridges.[57] The book aimed to attract readers with mathematical interests and to instill excitement about architecture as an intellectually stimulating subject, "neither wholly looking at old churches, nor laboriously calculating stresses in beams and loads with columns."[58] The authors contemplated uses of the book in the educational level of the sixth form—a year of advanced high school study that preceded university entry—as an introduction to modern ideas of architectural form and spatial organization.[59] In their preface, March and Steadman also stated alliances between structuralism and the new math, and the new math's promise to furnish architecture, amidst other qualitative fields, with a firm mathematical basis.[60] Doing so required bridging architects' knowledge gap between traditional concepts in geometry and arithmetic—the "old mathematics"—with the "new mathematics" of structures and relations.

In opening *The Geometry of Environment*, Leslie Martin declared mappings between architecture and mathematics to be natural, given the structural essence of both fields.[61] This was not about seeking out obscure formalisms for architecture. The book's images made isomorphisms seem undeniable, as if structures could be *found* in the environment. It simply required someone taught to recognize them.

The start of the 1960s found the topic of mathematical education in the UK acquiring a "feeling of . . . breathtaking urgency" for members of British schools, universities, and the industry.[64] An impactful conference on the teaching of mathematics that took place at the University of Southampton in April 1961 rang alarms for the dire national consequences of an unsuccessful mathematical education. These discussions paralleled debates in the United States that cast mathematics as a vehicle for scientific and technological supremacy. In Cold War America, governmental entities and the public put emphasis on

Interlude 2

This is a model of a school at about 1:50 scale. Or so one might speculate from the size of the metal clips that hold together the paper-thin walls. Black thread passes through every door, knots at eight intersections, and loops closed on the outside. Wood blocks, painted yellow and blue, lie in the center of five numbered rooms. A silver miniature car sits outside the main entrance.

This model was not made to design architecture, but to teach primary school children math.

It appeared in the 1969 textbook *Environmental Geometry* published by the Nuffield Mathematics Project, an educational initiative launched in 1964 to promote experiential and discovery-oriented approaches for children 5–13 years old.

The model, however, may have been made by architects. Most examples used in *Environmental Geometry* were produced by a group of first-year architecture students at the Bartlett School of Architecture led by faculty member George Kasabov, erstwhile spokesperson of the British Architectural Students Association.[62] The book featured photos of students building a house in a mathematics class or constructing pyramids and house plans with poles and ropes.

The paper and black-thread model was about teaching schoolchildren graph theory. Students would be asked things such as: "Which is the shortest route? Which is the longest route you could take without walking along the same part twice? Could you walk along all these paths without covering any part twice?"[63]

To answer these questions the paper model mattered little. The metal clips loosely holding it together succumbed under the power of the knots, and the white paper became mere background for the black thread: a hidden structure announcing its presence, yielding intellectual and calculative gains, and inculcating a new kind of seeing.

A graph vision.

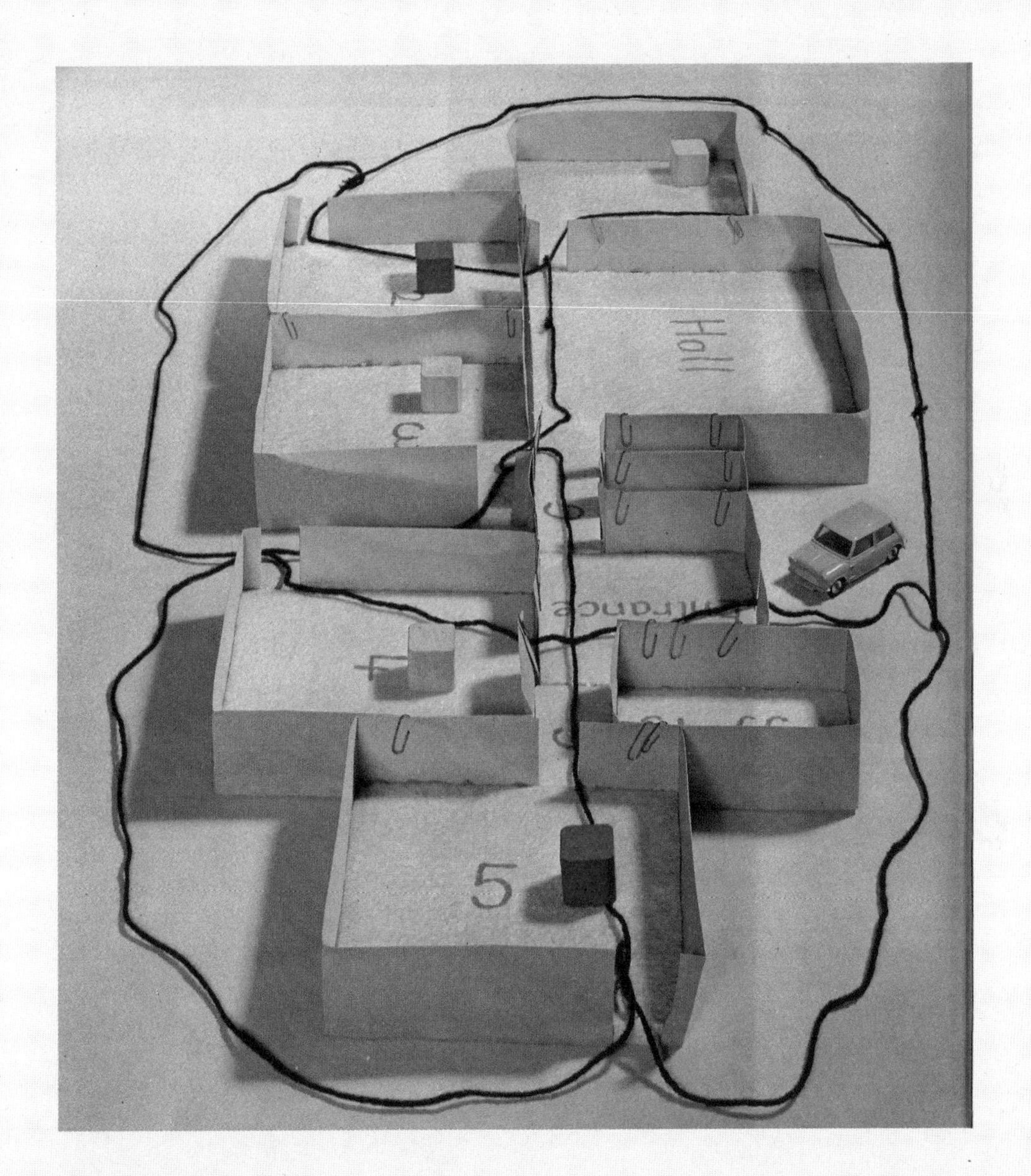

Figure 2.10
Source: Nuffield Mathematics Project, *Environmental Geometry* (London: Chambers and Murray, 1969), 38.

mathematical training, for providing both techniques and mental proclivities for deciphering patterns behind an increasingly complex and fast-changing world.[65]

In the UK, the Southampton Mathematical Conference was the culmination of conversations about problems and challenges in the teaching of mathematics, already under way since the late 1950s. Themed "Mathematical Models," the conference promoted models as a bridge between pure and applied mathematics.[66] Mathematical modeling was about identifying and interrelating variables and implicated processes of abstraction, mapping, and comparing—structural ideas that "pure" modern mathematicians also embraced. The conference laid the seeds of the most influential initiative of the British new math in secondary education, titled the "School Mathematics Project." Mathematician Bryan Thwaites cofounded the Project with the double agenda of evolving syllabi for ages 11 to 18 that would "adequately reflect the modern trends and usages of mathematics" and producing complete textbooks and teachers' guides.[67] Within ten years from the School Mathematics Project's establishment, its books were being used by around half the schools in England, giving it a leading position among many other organizations that engaged with the development of mathematical syllabi and textbooks in the UK throughout the 1960s.[68] School Mathematics Project textbooks familiarized students with mathematical topics other than the traditional arithmetic and geometry, such as sets, groups, matrices, transformations, and topology, while also promoting "structure and pattern" as intellectual ideals.

In his retrospective on the School Mathematics Project's first decade, deputy director Douglas Quadling declared the teaching of mathematics a social function to help students interpret and control their environment.[69] The increasing impact of "mathematical patterns of thought" on areas as diverse as industrial practice, ecological studies, urban planning, or "putting spacemen into orbit" mandated a teaching of mathematics aimed at what Quadling referred to as "appreciation," learning to *perceive* patterns and structures in the world.[70] The value of relating mathematical topics to the "living and thinking experience of the pupil" had been a matter of active debate since the Southampton Mathematical Conference and made it into the School Mathematics Project.[71] This set the Project apart from dominant intellectual traditions in modern mathematics that denounced the experiential world and proclaimed mathematics a predominantly intellectual activity. Quadling tellingly described *abstraction* "as a process in which the pupils should engage actively as a result of concrete experience, not as a system of laws to be imposed upon them."[72]

In the British new math's negotiations of abstraction and concrete experience, geometry was the subject of great debate and contention.

Figure 2.11
Cover of Book 5 of the School Mathematics Project found in the Martin Centre Library at the University of Cambridge. The computer paper tape motif featured on the covers reads "THE SCHOOL MATHEMATICS PROJECT DIRECTED BY BRYAN THWAITES."
Source: *School Mathematics Project: Book 5* (1969; Cambridge: Cambridge University Press, 1974). Reproduced with permission of the Licensor through PLSclear.

Figure 2.12
Front matter of the school mathematics textbook *Environmental Geometry* by the Nuffield Mathematics Project, initiated in 1964 to teach children ages 5–13 a contemporary approach to mathematics. *Source*: Nuffield Mathematics Project, *Environmental Geometry* (London: Chambers and Murray, 1969).

In his retrospective, for instance, Quadling described geometry as "the most difficult area of all" and voiced lingering doubts on whether geometry should have any place at all in the School Mathematics Project syllabi.[73] Such doubts go back to some of the origin stories of the new math debates, including infamous mottos such as "Euclid must go!" by prominent Bourbakist Jean Dieudonné.[74] Dieudonné dismissed geometric shapes as "artificial playthings" whose visual appearance was an excuse for sloppy reasoning.[75] The remedy he proposed was to expel visual reasoning from geometry and recast geometry as a kind of algebra.[76]

The School Mathematics Project was not as unswerving about replacing visual intuition and experience with axiomatic reasoning. Bodily experience and visual intuitions were seen as the basis upon which to present mathematical notions of sets, matrices, groups, and others in a spatial and experiential context. Thwaites wrote that in the place of a "watered-down Euclid" the School Mathematics Project strove to provide students with "a feeling for . . . spatial relationships" and to emphasize the interplay between geometry and algebra.[77] A(dvanced)-level students were being exposed to more abstract, algebraic renditions of geometry, after having first passed through the general O(rdinary)-level course that emphasized ideas such as the topology of geometrical figures, patterns, transformations, and visualization.[78]

The School Mathematics Project's approach to geometry was influenced by late nineteenth-century German mathematician Felix Klein, often positioned at Hilbert's antipode as an "anti-modern" because of his support of mathematics' interconnection with the experiential world.[79] Klein contended that mathematical elements carried meanings outside the realms of formal definition and that mathematics progressed as much with logic and reasoning as it did with perception and imagination. He advanced a new philosophy of geometry that came to be known as the Erlangen Program, from a famous lecture that he delivered at the Philosophical Faculty of the University of Erlangen in October 1872 upon being appointed full professor.[80] The Program maintained figures (shapes) at the center and used the properties of these figures that remained unchanged (invariant) under certain transformations to define different geometries, in particular bringing Euclidean and non-Euclidean geometry under the umbrella of projective geometry.[81] Quadling's retrospective cast transformations as offering a propitious future for geometry within the new math pedagogy:

> The approach through transformations, which had been in the background since Klein's lecture in 1872 . . . seemed to offer new hope. In its early stages it gives much scope for practical activity based on children's experience of movement and aimed at building

up spatial awareness. As the work develops, we find central themes of mathematics—sets of points, transformations (or functions), groups, etc.—exhibited in a spatial context.[82]

Indeed, the School Mathematics Project O-level textbooks were amply illustrated with examples showing mathematical concepts of mappings, transformations, and topological diagrams applied to familiar things such as house plans, electrical networks, and geography maps. This approach was also adopted at the elementary school level, where, as we have seen, the Nuffield Mathematics Project's felicitously titled textbook *Environmental Geometry* deployed floor-plan drawings, models, and house-building exercises to introduce students to geometry and topology.[83] The imagery presented in these textbooks tapped into students' experience and intuitions to foster understanding of mathematical concepts and provided students with a mathematical lens for seeing the world. In other words, it not only relied on spatial intuitions but also shaped them.

The Geometry of Environment was written in the context of the new math textbooks. For one, it overlapped in terms of topics and method of illustration—new mathematical ideas illustrated through examples from the built surroundings. More directly, new math textbooks and the broader intellectual context of the new math reforms in the UK cultivated its authors' mathematical dexterities. Although there was no explicit requirement that LUBFS Centre researchers should be mathematically proficient, the Centre attracted architecture students with an "abstract mind and an interest in theoretical questions" and who were "naturally inclined to quantification and rational thinking."[84] These were inclinations mostly developed at the high school level. Several of the Centre's members entered architecture with ample classroom training in the new mathematics of relations and structures. Steadman, in fact, reports being Thwaites's student and acknowledges that his high school experiences shaped his ideas about how mathematics might be applied in architecture.[85] For March, the school exposure to the kinds of mathematics later consolidated as the "new maths" was what pushed him toward mathematical studies in Cambridge and, one year later, away from them, to pursue stage design and ultimately architecture. March attributed his change of heart to an "immense boredom" with university mathematics, which was high school "all over again."[86]

Graph theory was not a central subject in new math textbooks, but shared intimate links with two of their core topics: set theory and matrix algebra. A graph could be represented as a table of numbers, a matrix, while several set theoretic concepts could be translated into graph

Exercise E

1. State which, if any, of the line sketches in Figure 34 are topologically equivalent:

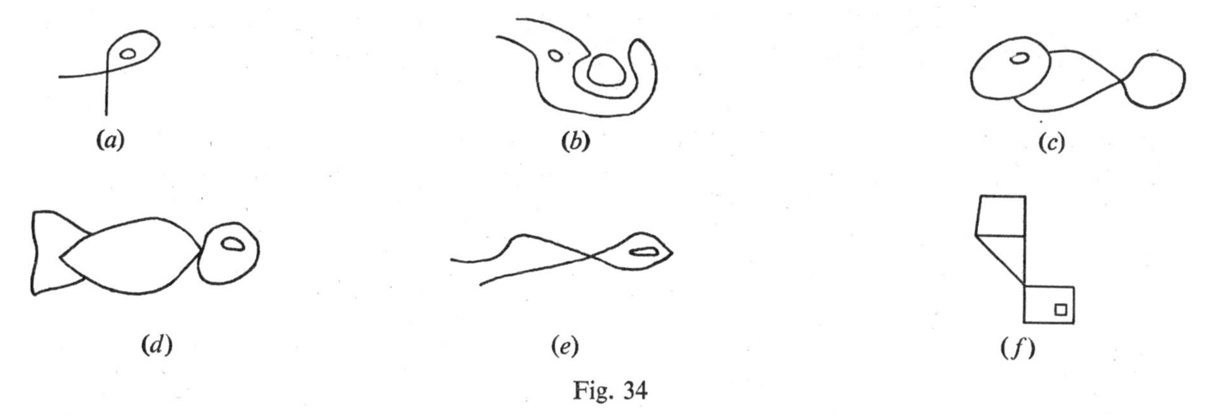

2. Pick out any of the line sketches which in Figure 35 are topologically equivalent to each other:

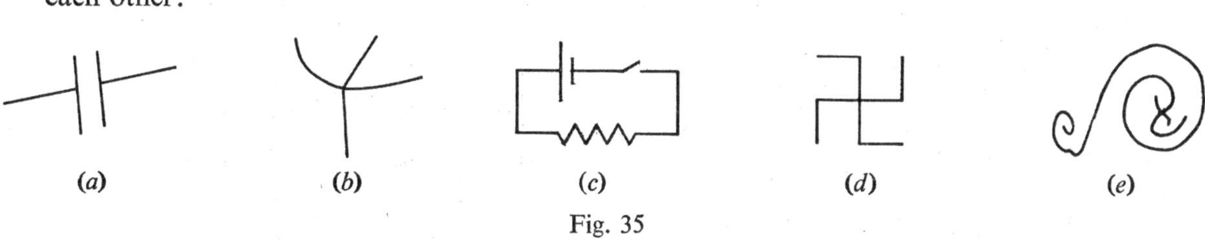

Figure 2.13
Exercise on invariants and transformations from the chapter on geometry of School Mathematics Project Book 5. The exercise asked students to identify topological equivalences between dissimilar forms blatantly severed from their symbolism. *Source*: *School Mathematics Project: Book 5* (1969; Cambridge: Cambridge University Press, 1974), 261. Reproduced with permission of the Licensor through PLSclear.

theoretic terms and vice versa. Steadman learned graph theory through self-instruction, a process encouraged in the UK new math culture.[87]

Far from being a marginal topic for the mathematically curious, graph theory encountered a true surge of interest in the 1960s. Until then, graph theory had been seen as trivial and unrigorous, "the so-called science of trivial and amusing problems for children, problems about drawing a geometrical figure in a single sweep of the pencil, problems about threading mazes, and problems about collating maps and cubes in cute and crazy ways."[88] Blanche Descartes's Bill Tutte recalls that students "tempted" by the subject were advised "to turn to something respectable or even useful, like differential equations."[89] By the end of the 1960s, the situation had shifted, with graph theory textbooks proliferating and applications of graph theory spreading across disciplines. Graph theory's infamy gave way to celebration, even hyperbole. For example, the 1972 English translation of a 1968 French textbook on graph theory ended with an elegy on graphs as heralds of humanism:

Figure 2.14
Photographs of children's constructions with poles and ropes, including a pyramid and a house plan. *Source*: Nuffield Mathematics Project, *Environmental Geometry* (London: Chambers and Murray, 1969), 41.

> No engineer, or physicist, or chemist can ignore this theory; otherwise, complicated structures will become for them labyrinths which lie outside scientific method. . . . The engineer and the artist will be brought close together by the theory of graphs and its extensions; will it not be a good thing for them to join together in logical and global understanding and thus be more completely human?[90]

Such praise was common among mathematicians who promoted graph theory by emphasizing its novelty, its accessibility, and its potential in delivering one of the key intellectual mandates of the period: the unification of all sciences under structural descriptions. Mathematician Frank Harary, for example, tirelessly preached the usefulness of graph theory for practically every field of human knowledge. He argued that graphs were prime models of empirical phenomena. "When these terms [graphs]," he wrote, "are given concrete referents, digraphs serve as mathematical models of empirical structures, and properties of digraphs reflect structural properties of the empirical world."[91] Some twenty years later, he put this statement to prose and posited that "for every field that has a structure, graph theory yields material for a lecture."[92]

Graphs enjoyed vast popularity in disciplines that turned to structuralism in search of scientific rigor. Persuasive preachers, such as Harary or Øystein Ore, capitalized on the cultural relevance of structuralism to highlight their benefits. Graphs remained faithful to the tenet of studying relations instead of properties of objects, an idea carrying cultural connotations of modernization and unification. Yet, as Harary incessantly argued, their geometric appearance offered advantages of accessibility and congeniality that were missing from, for example, set theory or matrix algebra. Graphs rendered these topics—pinnacles of structural studies—visible, applicable, concrete, and teachable. It gave structures and systems a workable image. Especially for architecture, the oscillation of graphs between the empirical and the abstract fostered hopes of instilling a new form of representation that had a visual appearance but remained immune to the treacherous ambiguity and seductiveness of geometric shapes.

We may now begin to see *The Geometry of Environment* in a different light. The book did not only update architects on the mathematics proliferating in British schools and fueling high science and technology; it carried with it lively debates and enticing possibilities within the new math culture. In opening the first chapter of the book on "Mappings and Transformations," March contended that architectural orthographic drawings were, in essence, *maps*. They matched points in physical space to points in a drawing in a one-to-one way.[93] Architectural drawings themselves could be mapped onto each other through

Figure 2.15 (*opposite*)
Use of a famous architectural example (Le Corbusier's Maison Minimum) to explain the mathematical concept of mappings. From top to bottom the Maison Minimum is subjected to mappings of identity, isometry, similarity, affinity, perspectivity, and topology. *Source*: Lionel March and Philip Steadman, *The Geometry of Environment: An Introduction to Spatial Organization in Design* (Cambridge, MA: MIT Press, 1974), 20. Copyright Philip Steadman and the Estate of Lionel March.

various transformations—the pinnacle of Klein's Erlangen Program, which as we saw shaped the School Mathematics Project approach to geometry. For example, March noted, inking in a pencil drawing of an elevation was the "identity" transformation of the elevation onto itself. Transformations that preserved lengths but allowed for positional changes were "isometries," while those of scaling a drawing up and down were "similarities." Transformations of parallel lines onto parallel lines were "affinities," while those that changed angles and lengths were "perspectivities" (a special case of "projectivities"). Finally, transformations that preserved neighboring relationships between points were "topological" transformations.[94]

Unlike affinities, isometries, and perspectivities, topological mappings dispelled any intuitive sense of correspondence between two images, while also working to declare images that were strikingly dissimilar as equivalent. For instance, March presented a semidetached version of Le Corbusier's Maison Minimum, which he announced as topologically equivalent to the detached house.[95] A few pages later, the three unbuilt Frank Lloyd Wright houses we saw in Interlude 1 were shown to map onto the same graph, revealing structural sameness under their wildly divergent geometric languages. "Sometimes," March wrote right before the reader turned to the page where a graph uttered the surprising revelation, "objects which appear to be very dissimilar on first acquaintance may be seen, later, to share an underlying structural pattern."[96]

The intellectual implications of a new way of seeing architecture through structure rather than appearance echoed strongly in chapters 10 and 11 on graphs, both written by Steadman. The chapters featured orthographic representations of well-known architectural projects alongside set and graph models of these projects. This side-by-side juxtaposition was not a claim for a complete replacement of architectural drawing by mathematical models. It was, however, a clear subjugation of the surface vision of geometric representations to abstract structural representations. Such was the graph's value, its capacity to "show the essential structure of a set of relationships."[97]

In chapter 10, "Planar Graphs and Relations," Steadman introduced readers to the "adjacency graph," a graph whose points represented separate rooms and whose lines represented connections between rooms. He pointed out that the plan *itself* could be regarded as a graph. The graph's lines would represent walls or boundaries and the points would represent their intersections.[98] These two graphs were linked to each other with the "dual" relationship. A "dual" of a graph A is another graph B that has a point for each area in A bounded by lines. The "dual" relationship between the floor plan and the adjacency graph suggested that a direct mapping could be made between the

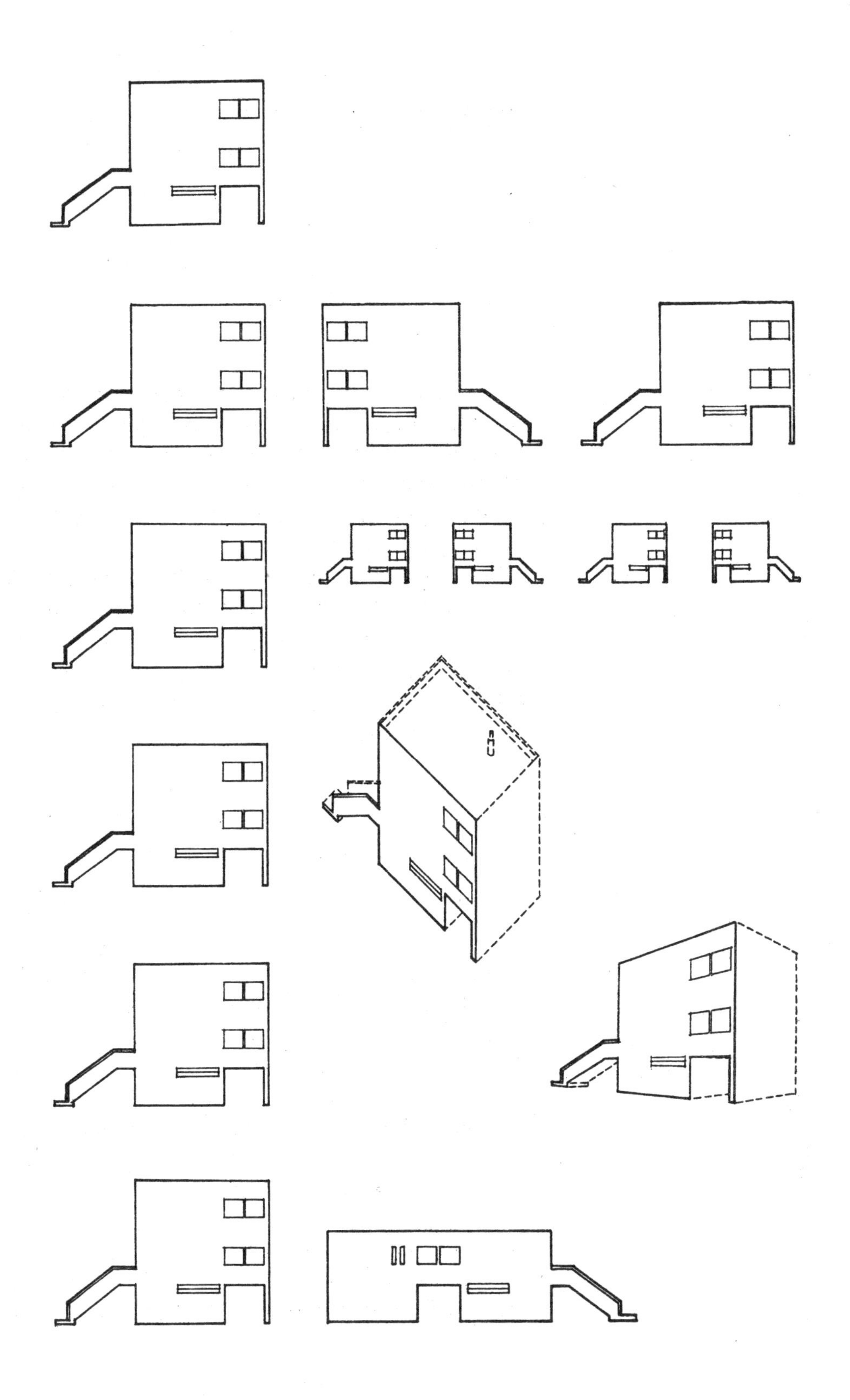

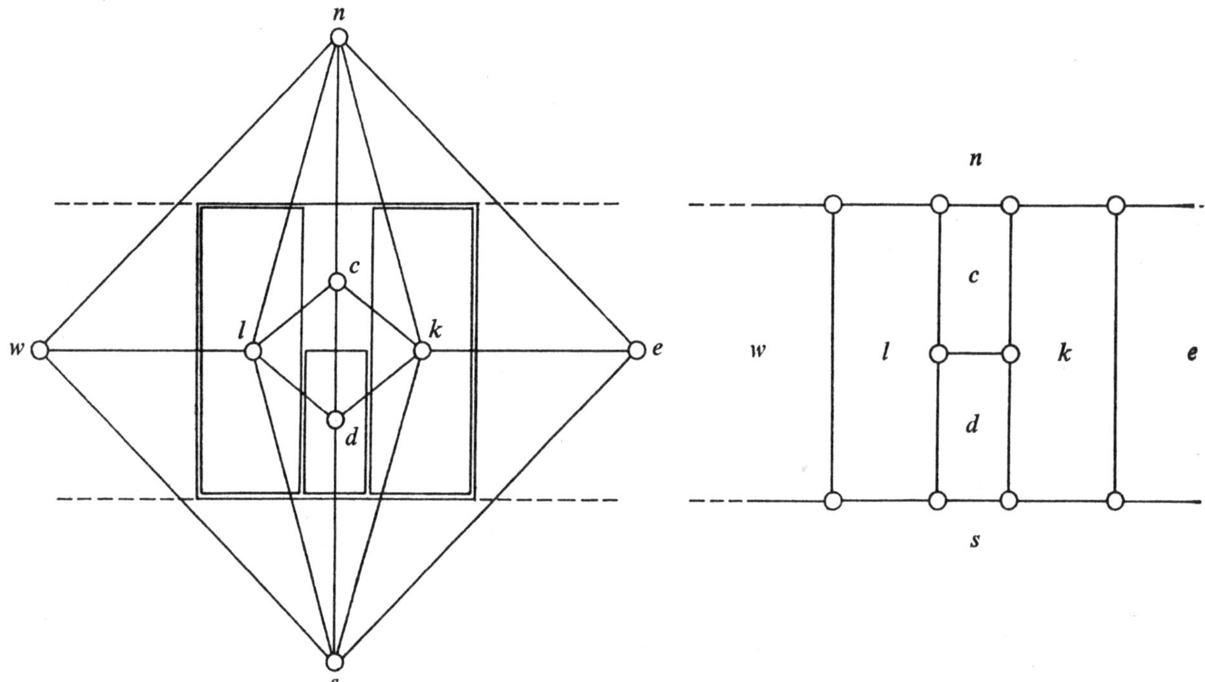

Figure 2.16
A house plan represented as a graph (right); the house plan's graph is the "dual" of the "adjacency graph" (left), which represents the architectural program (types of rooms and connections between rooms). *Source*: Lionel March and Philip Steadman, *The Geometry of Environment: An Introduction to Spatial Organization in Design* (Cambridge, MA: MIT Press, 1974), 254. Copyright Philip Steadman and the Estate of Lionel March.

description of an architectural program and the geometry of an architectural floor plan—form and function mapped onto each other.

Later in the chapter, the reader was presented with plan views of Frank Lloyd Wright's Aline Devin House (1896) in Chicago. Wright's project was renowned for the ingenious way in which the architect had managed to preserve the independence of rooms within a symmetrical plan—a requirement, however, that had pushed Wright to a two-story arrangement. Steadman put Wright's work in the context of Viollet-le-Duc's discussion of the changes in "house-planning" in post-Revolutionary France and the need for the replacement of the erstwhile social distance between ruler and servants with physical barriers to ensure the privacy of different families. "This new *individualism* in society," Steadman observed, "implied isolation and the distinction of parts in the dwelling."[99] Such consideration of the relationships between different rooms and circulation paths, Steadman suggested, motivated by Wright's commitment to individualism and democracy, was the driver behind the Devin House design.

Graphs placed these decisions under scrutiny. Reformulating, understanding, and potentially revising foundational works and historical debates in the architectural discipline was, after all, one of the book's key tropes. Steadman showed plans of the Devin house's two levels redrawn in the book's neat graphic style, stripped of Wright's original wall *poché*, and with an index of room types indicated by letter labels on the plan. A 45-degree axonometric cut at the longitudinal axis with dramatic black fill figures on the next page. The reader was then presented with a drawing that at first glance looked like a diagram of an electric circuit and which the label identified as an "adjacency graph." 3 mm circles indicated the rooms, labeled with letters, and lines indicated their connections. Jagged lines showed level changes, and a dashed line connected the garden with the terrace. Steadman explained that the purpose of the project's graph theoretic re-representation was to ask the question "does the structure of the problem require such a complicated solution" instead of "sitting back and enjoying the solution, in the same way that the architect clearly enjoyed at arriving at it."[100]

Steadman applied a theorem by Polish mathematician Kazimierz Kuratowski for identifying whether a graph is planar, that is, whether it can be embedded in a plane. To ensure planarity, one needs to examine the adjacency graph and rule out the presence of two "forbidden" subgraphs, a graph of six points and nine lines known as the "utilities graph" and the pentagonal graph consisting of five points completely linked.[101] Planarity meant that an adjacency graph could be resolved as a floor plan in a single level only. Inspection of the Devin house's adjacency graph declared it nonplanar. The reason, however, Steadman diagnosed, was the treatment of two alleyways, one giving access to the cellar and one to the boiler room, as separate nodes in the graph even though they connected through the garden. The culprit was *aesthetic* rather than functional; it was Wright's adherence to bilateral symmetry that compelled him to separate the alleyways. Steadman continued to speculate:

> It can hardly be the intention of the plan, though, that the coalman should pass along this route, across the lake-shore panorama in full view of the terrace and garden, only to meet the wine merchant coming around the path in the other direction. . . . In any event the requirement that access to a cellar and to a boiler-room should be kept separate is hardly the stuff with which to wage an architectural polemic for democracy, individualism, and Republicanism![102]

Combine the two alleyways, and the new adjacency graph *is* planar. Steadman did not venture to provide orthographic drawings of this emergent other Devin House (perhaps this was left as a challenge to the architecture readers of the book) but anticipated that its plan would not be symmetrical. The verdict was, however, that "Wright's youthful architectural dexterity is certainly fun, but is *not* functionally necessary."[103]

The next chapter in *The Geometry of Environment* on "Electrical Networks and Mosaics of Rectangles" focused on *networks*, which Steadman defined as graphs whose lines have a direction (directed graphs)

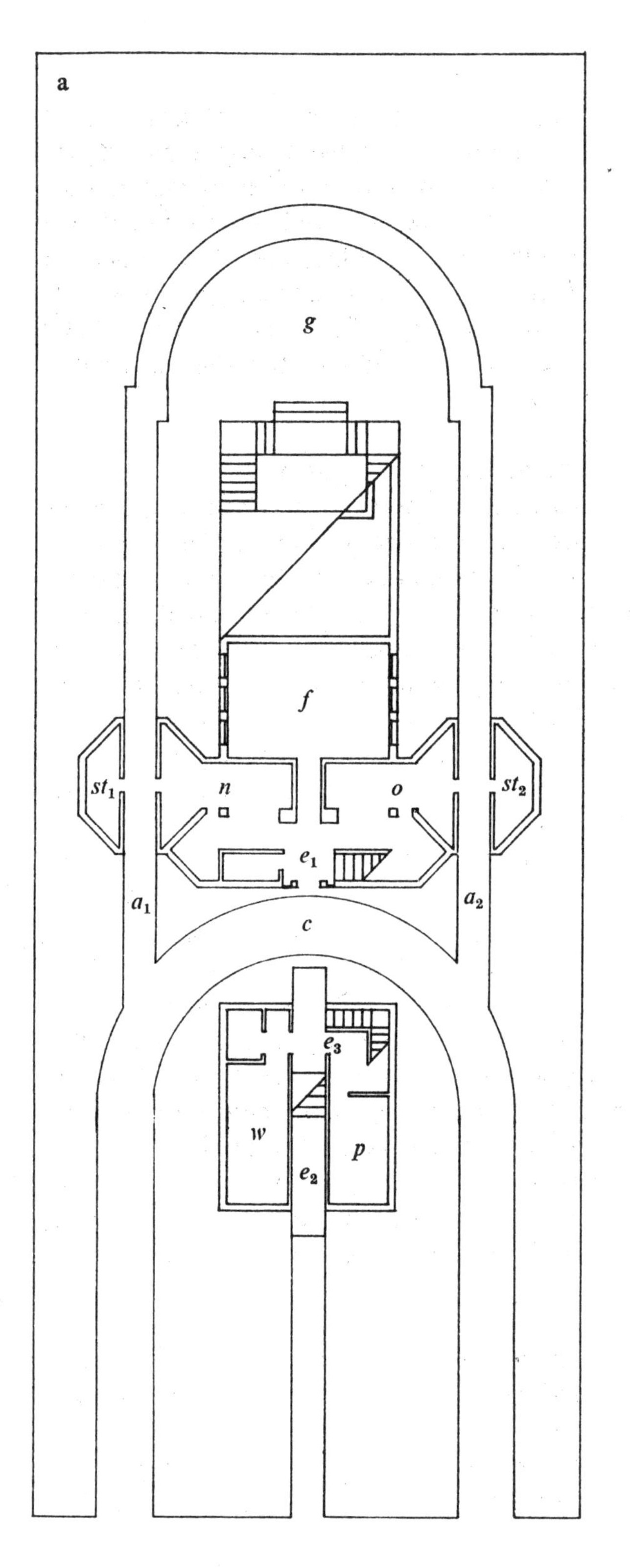
a
g
f
st_1
n
o
st_2
e_1
a_1
a_2
c
e_3
w
p
e_2

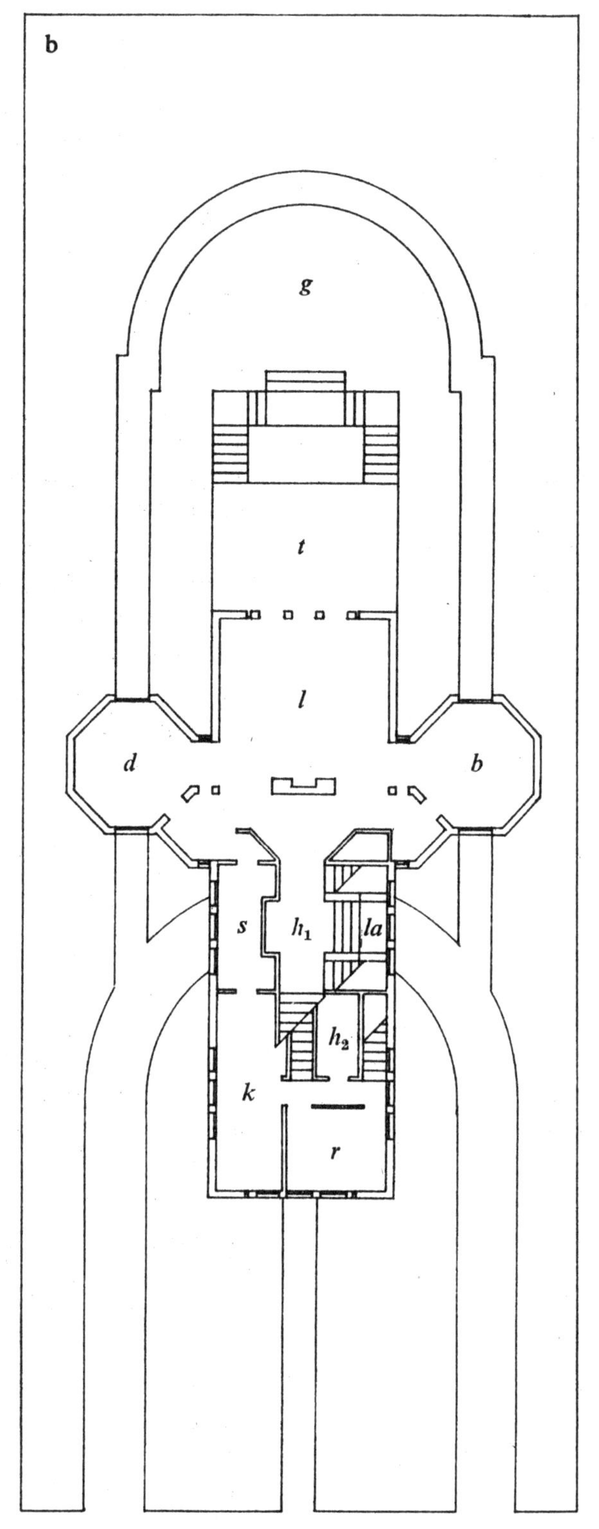
b
g
t
l
d
b
s
h_1
la
h_2
k
r

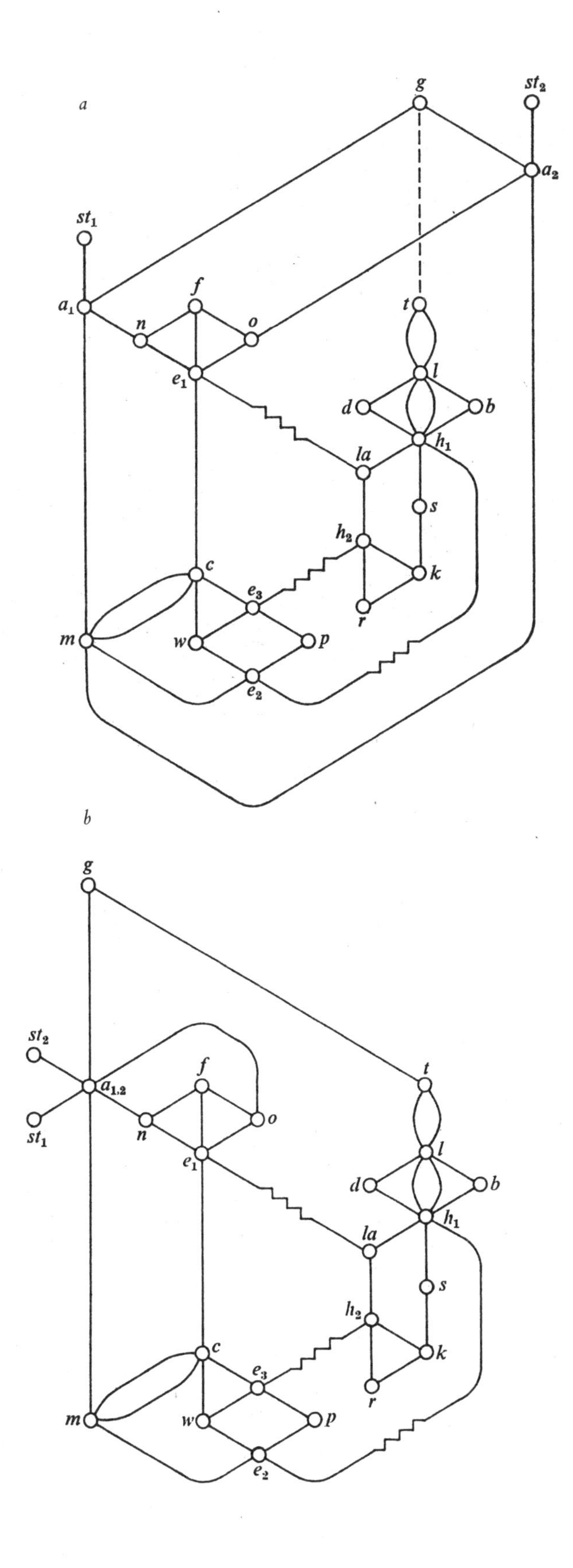

Figure 2.17 (*opposite*)
Floor plans of Frank Lloyd Wright's Devin House in Chicago (1896) revealing the building's intricate circulation system. *Source*: Lionel March and Philip Steadman, *The Geometry of Environment: An Introduction to Spatial Organization in Design* (Cambridge, MA: MIT Press, 1974), 258. Copyright Philip Steadman and the Estate of Lionel March.

Figure 2.18
The Devin House represented by its corresponding adjacency graph (rooms and connections between them). The graph is not planar and therefore cannot be resolved in a single floor layer (*a*). A new adjacency graph merging the alleyways a1 and a2 makes the graph planar and renders a second floor unnecessary (*b*). *Source*: Lionel March and Philip Steadman, *The Geometry of Environment: An Introduction to Spatial Organization in Design* (Cambridge, MA: MIT Press, 1974), 261 (*a*), 262 (*b*). Copyright Philip Steadman and the Estate of Lionel March.

and are assigned a numerical value.[104] Aligned with governmental pursuits of standardization and minimum housing, Steadman used graphs to fit rooms of specific dimensions within a given boundary while preserving a desired topological relationship of adjacencies and proximities among these rooms.[105] The material for the chapter was taken from LUBFS Centre Working Paper 23, "The Automatic Generation of Minimum-Standard House Plans," that Steadman wrote in 1970.[106] In the late 1960s, the National Building Agency (NBA) had published a "Generic Plans" catalog and one of "Metric House Shells," which listed 159 house shells that allowed for 400 distinct possible plan types.[107] These catalogs were based on minimum dimensional standards developed by the Parker Morris Committee and presented in their 1961 Report *Homes for Today and Tomorrow*, and on activity requirements from the 1968 Ministry of Housing and Local Government pamphlet *Space in the Home*.[108] The National Building Agency's aim, which Steadman approached with a mathematical spirit, was to adopt a cost-effective set of dimensional standards for industrialized building construction, "without any significant reduction of choice in layout or design."[109]

To produce such dimensional alternatives, Steadman applied a method for dissecting rectangles into squares developed in 1937 by the "Trinity Four," Tutte, Brooks, Smith, and Stone—our Blanche. Often referred to as "squaring the square," the method was popularized in the 1960s by mathematics and science writer Martin Gardner.[110] The method included what Steadman referred to as a "curious analogy" between the problem of fitting rectangles within a given rectangular boundary (that is, rooms within a given shell) and the flow of electricity in a closed circuit.[111] When represented as directed graphs, both electrical circuits and architectural floor plans were governed by the same rules. In the electrical circuit there are amounts of current, while in the floor plan there are sizes of rooms. In both cases, the amount (current or size) entering a node of the graph needs to equal the amount leaving that node. This analogy would make it possible to calculate how rooms of specific sizes could be distributed within a shell of given dimension. Steadman's graphs stipulated Kirchhoff's second law to be true for architectural floor planning.

Aside from revealing unlikely analogies between architecture and physics, the graph mapping of an architectural floor plan was also amenable to computer implementations. Although computers were not yet readily available at Cambridge—the Cambridge Mathematical Laboratory only had one computer, the Titan, an epigone of the experimental EDSACs—there was work being done in translating mathematical models into "data structures."[112] This argument did not make it into *The Geometry of Environment* but was an important part

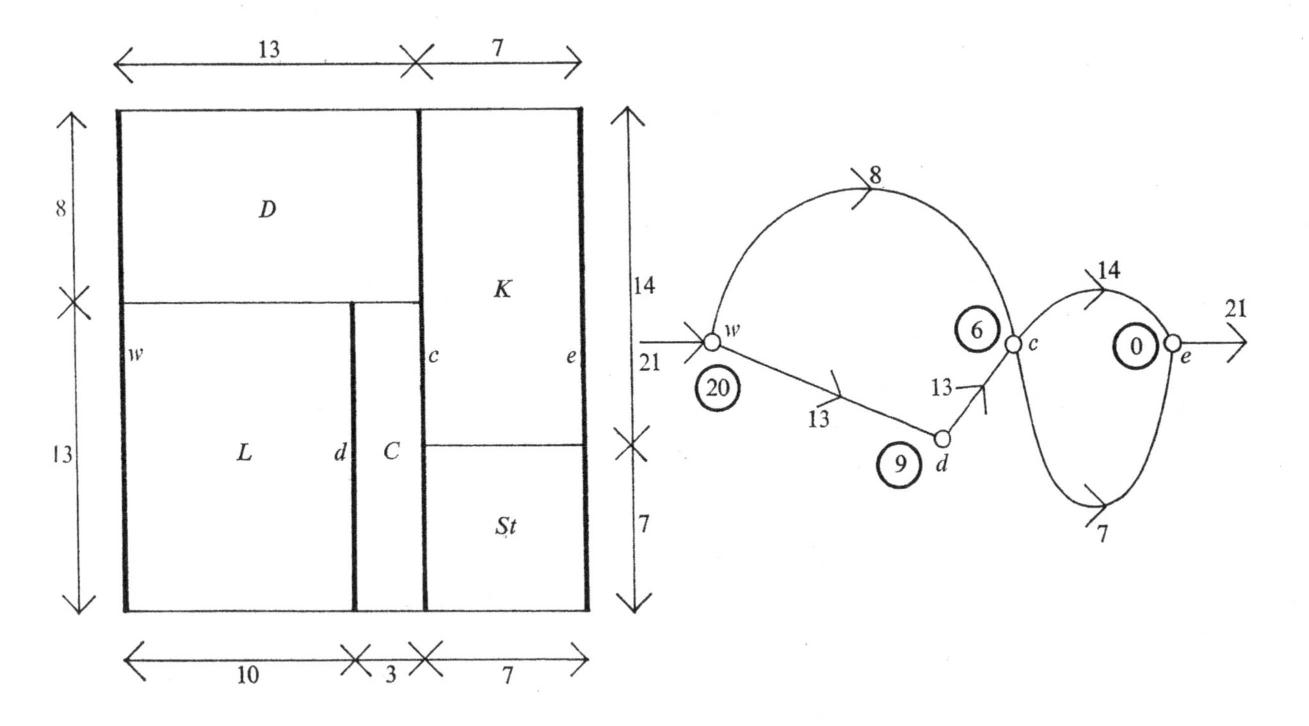

Figure 2.19
A digraph representing the dimensions of the vertical walls in the schematic floor plan on the left. *Source*: Lionel March and Philip Steadman, *The Geometry of Environment: An Introduction to Spatial Organization in Design* (Cambridge, MA: MIT Press, 1974), 275. Copyright Philip Steadman and the Estate of Lionel March.

of Working Paper 23. In the paper, Steadman referenced Cambridge mathematician and computer scientist Crispin Gray. He explained that data structures were also graphlike: they consisted of data-containing blocks, interconnected by relation-representing pointers to be optimally efficient for specific transformations of a mathematical model. The relational structure of the graph was easily representable in a machine by an "incidence matrix": a tabular equivalent of a graph representing the graph's vertices and their connections in binary form. In discussing the transcription from graph to computer code Steadman appeared wary of conflicts between the dynamic nature of the graph computation done by hand, which he described in his paper, and the static representation of the graph in a computer. The nodes, he explained, needed to be updated dynamically.

This double nature of the graph as *both* a visual entity that could be intuitively grasped in a singular act of seeing *and* a table of numbers ("incidence matrix") was attractive. The graph held together numbers and image, reason and vision, rational abstraction and empirical concreteness, offering a hopeful bridge between architecture's long disciplinary traditions preoccupied with seeing and the forceful intellectual mandate of structural abstraction that swept through academic departments in the 1960s. For all the claims of modernization that accompanied them, architectural engagements with graphs were not forward projective but revisionist. They were about *seeing* past works

Figure 2.20 (*opposite*)
Collage (*a*) and legend (*b*) of members at the Martin Centre for Architectural and Urban Studies (formerly Land Use Built Form Studies Centre) in the style of the Beatles' *Sgt. Pepper's Lonely Hearts Club Band* album cover. The collage was made in 1992 by Andrew Coburn. *Source*: Philip Steadman personal collection.

of architecture under a different light, a "hard intellectual" one. The break with the past was not through a renewal of architectural languages, meanings, tectonic strategies, but rather an overhaul of architectural visualization through the making of images that could be argued *not* to be images. This oxymoron's allure extended beyond the confines of university research centers. It found unlikely interlocutors among leading figures in "radical" or "experimental" architecture for whom graphs offered an opportunity to negotiate alliances with a mathematically inclined architectural intelligentsia while preserving the aesthetic specificity of their work.

Frameworks

At the apogee of his fame for fashioning dazzling images of a near-future utopian world where gridlike ever-extending space frames afforded unhindered spatial and social mobility, Hungarian-born French visionary architect Yona Friedman decided that images would no longer do. In the *mise au point* (perspective) of his 1971 book *Pour une architecture scientifique*—the title likely a playful adaptation of Le Corbusier's fabled *Vers une architecture*—Friedman lamented that though the formal ideas of his work were being widely imitated, their theoretical core was poorly understood. Forward-thinking architects around the world such as Eckhard Schulze-Fielitz, Kenzo Tange, Kisho Kurokawa, the Archigram group, Moshe Safdie, Ricardo Bofill, and Erwin Mühlestein had variously adopted the ideas of mobility, modularity, and three-dimensional urbanism that he was promoting, but mostly failed to appreciate the reasoning behind them. *Pour une architecture scientifique* intended to remedy this situation by presenting, as Friedman wrote, "nothing but reasoning, without wandering off into visualizations of the results which this reasoning has produced."[113]

Despite Friedman's programmatic rejection of visualization, the book was heavily illustrated. The 63 hand-drawn figures that appeared in the main body of the book were, however, of a different kind. The familiar imagery of inhabitable megastructures hovering over the English Channel, the Place de la Concorde, or Frank Lloyd Wright's Imperial Hotel in Tokyo as counterargument to its demolition were absent.[114] Instead, the book was populated by drawings of graphs. Friedman placed graphs alongside sketches of plan layouts, schematic drawings of space frames, and charts of an individual's living habits. Graphs figured in the book as the thread that tied together all aspects of Friedman's theoretical articulations and architectural experiments since his impactful entry into the international architectural scene in the mid-1950s.

1 Bill Mitchell
2 Alex Pike
3 Tony Shadbolt
4 Eric Dudley
5 Simon Ruffle
6 Stephen Platt
7 François Penz
8 William Fawcett
9 Tim Lewers
10 Andy Coburn
11 Helen Mulligan
12 Koen Steemers
13 Colin St John Wilson
14 Anna Dickens
15 Peter Carolin
16 Phil Tabor
17 Ola Uduku
18 Richard Baxter
19 Leslie Matthews
20 Dagmar Motycka
21 Brian Ford
22 Raf Orlowski
23 Mike Barron
24 Czaba Deak
25 Stan Laurel
26 Doug Cawthorne
27 Oliver Hardy
28 David Crowther
29 Tim Wiegand
30 Simon Schofield
31 John Rees
32 Monica Andrade
33 Phil Steadman
34 Richard Stibbs
35 On Original Cover
36 Catherine Cooke
37 Lee-Jong Lee
38 Ed Hoskins
39 Antonios Pomonis
40 Ian Cooper
41 Janet Owers
42 Christine Woodhouse
43 Marion Houston
44 Maria Sylvester
45 Marcial Echenique
46 Nick Bullock
47 Bill Howell
48 Lionel March
49 Robin Spence
50 Nick Baker
51 Paul Richens
52 Dean Hawkes
53 Sir Leslie Martin
54 An Apple Macintosh SE/30
55 A Cambridge Architectural Research Sweatshirt
56 Croquet mallets and hoops
57 Cambridge College buildings in the form of the Martin Centre logo
58 The Martin Centre Cat

NOTES
This picture was made on an Apple Macintosh Quadra computer using a Canon colour scanner/printer with Adobe PhotoShop software by Andy Coburn of Cambridge Architectural Research Limited. We gratefully acknowledge the assistance of the Martin Centre CADLAB. We apologise to the many who could not be included in this picture because we could not find photographs in the short time available.

Figure 2.21
Spatial city (n.d., binder of 1960s), photomontage by Yona Friedman on Lucien Hervé's photographs of the Imperial Hotel Tokyo, designed by Frank Lloyd Wright. *Source*: Fond Denise et Yona Friedman. Folder 294, page 25. Courtesy Fond Denise et Yona Friedman, Archives photographiques Jean-Baptiste Decavèle, Marianne Friedman Polonsky, https://www.yonafriedman.org/. All rights reserved.

In 1956 Friedman traveled from Haifa, where he was working as an architect, to Dubrovnik to participate in the 10th CIAM. In the meeting that came to mark CIAM's end, Friedman first voiced his critique of the functional determinism advanced in the Congresses. A pamphlet that he circulated at CIAM criticized architects' use of "pseudo-theories" to enforce their own values and ideas about architecture, thus defining and oppressing the lives of future users. The pamphlet put forward ideas of modular temporary dwelling and self-construction.[115] Soon after Dubrovnik, Friedman published his manifesto, in the form of two cyclostyled editions two years apart consisting of 300 copies each, for a "mobile architecture," a theory "stemming from the public domain" and accommodating "all personal hypotheses." He also circulated the principles of his mobile architecture in articles published in European journals—the earliest being his pre-CIAM 10 presentation of the idea in the German journal *Bauwelt*.

One year later, having relocated from Haifa to Paris, Friedman founded the Groupe d'Études d'Architecture Mobile (GEAM). The GEAM included like-minded architects David Georges Emmerich, Werner Ruhnau, Günter Günschel, Frei Otto, Paul Maymont, Eckhard Schulze-Fielitz, and others, who published manifestos bringing together ideas of transformability, change, and participation and bundled them with the unmistakable architectural language of space frames and modular structures. From 1958 to 1963 Friedman developed and avidly promoted the architectural expression of his theories, the so-called Ville Spatiale, through publications, exhibitions, teaching, and lectures internationally. The Ville Spatiale was a systematic megagrid supported by pillars 200–250 ft apart, which contained circulation, electric, and water installations. The space frame contained voids of 300–400 sq ft that the inhabitants could fill up to halfway with ephemeral "dwellings or offices," as Friedman specified, positioned in an unconstrained manner.

Pour une architecture scientifique presented the theory behind the Ville Spatiale form, or rather the theory that made the form mathematically *necessary*. The Ville Spatiale, Friedman seemed to argue, was no expression of high technology or the mobilities, unpredictabilities, and fluidities of rising neoliberal capitalism, which animated the work of some of his contemporaries and have become a steady interpretive framework for historians looking at 1960s "experimental architecture."[116] The Ville Spatiale, instead, was a mathematically founded device for rendering possible a form of social betterment. Friedman's work, anchored in arguments around the ethics of decisions and the inhabitants' right to choose and change their living settings, had strong political connotations, to which we will return. As potent as its bold social vision were the book's patent intellectual allegiances. Although

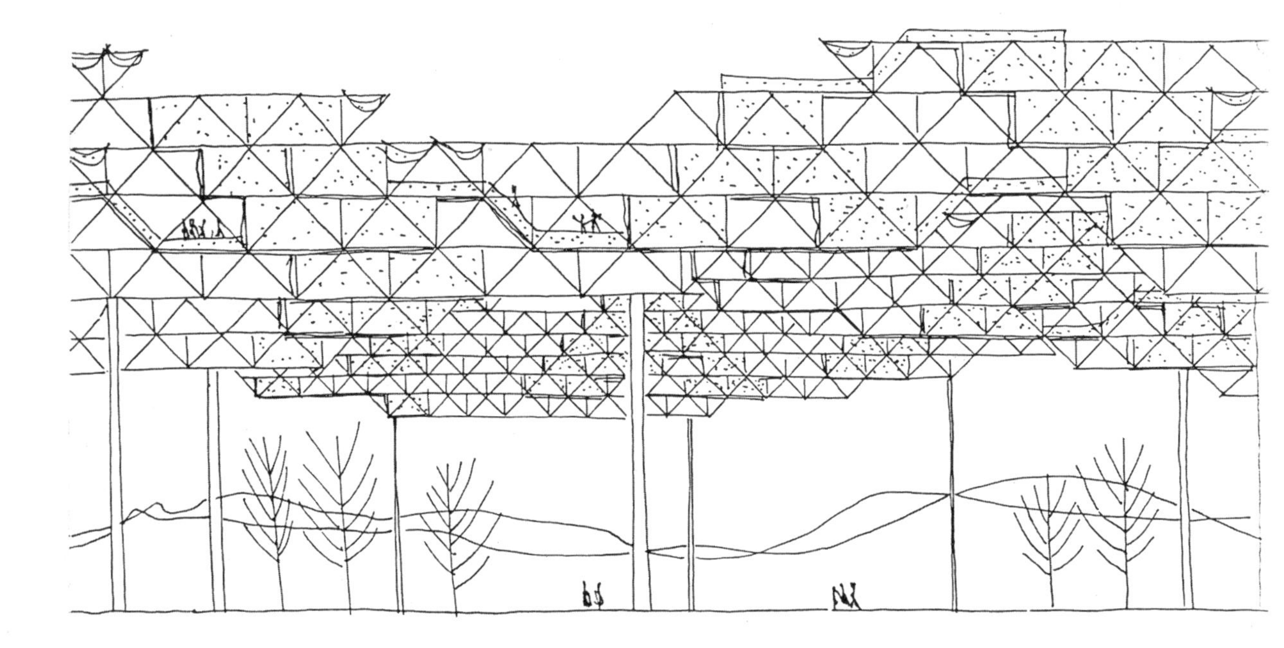

Figure 2.22
Ville Spatiale (1959) by Yona Friedman. *Source*: Fond Denise et Yona Friedman. Folder 303, page 252. Courtesy Fond Denise et Yona Friedman, Archives photographiques Jean-Baptiste Decavèle, Marianne Friedman Polonsky, https://www.yonafriedman.org/.

Pour une architecture scientifique did not include references, the book presents conspicuous alignments with modern mathematical culture and with structuralism. Friedman's book aspired to articulate an "objective," "axiomatic" theory of architecture that could form the *substructure* of all idioms and personal meanings.[117]

The image of that theory was the graph through and through. Early in the book, Friedman used a planar directed graph to reason about the attributes of the theory itself. Friedman took the graph's points to represent the theory's statements and the directions of the lines the order by which these statements were sequenced.[118] He then used this representation to reason on a topology for his theory that would guarantee it to be consistent (no contradictions among statements), nonredundant (each statement used only once), and complete (no "unenunciated" external statements). The graph showed the correct number of axioms for a theory of architecture and planning to be three. Connecting three statements was straightforward; there was only one way to connect them (they made a triangle), and sequencing them was easy. It sufficed to use a rule that wanted one arrow "entering" a point and one "leaving" it. Four statements made finding a correct order more difficult, and five (or above) resulted in crossings between the graph's lines (the graph was no longer planar). Crossings created a new relationship between statements that needed another implicit statement to be resolved, violating Friedman's last tenet of "completeness."[119]

Strikingly allowing the graph to vacillate between a model of an axiomatic theory and a familiar model of architectural floor plans,

Friedman contended that all of architecture and planning could be captured in the following three statements: "1. [architects and planners] make enclosures in pre-existing space; 2. For each enclosure there is at least one path leading to every other enclosure; 3. There are at least two different kinds of enclosures."[120] The graph's points, Friedman argued, had a "one-to-one" correspondence with the real spatial structure of an architectural plan, thus enabling immutable translations between reality and its representation.[121] Friedman next presented a schematic floor plan that consisted of three rooms, its adjacency graph, and its adjacency matrix—a table of numbers representing the graph in numerical form.[122] The graph's labels, indicating shape, equipment, and position of spaces, introduced considerations about context—"country, social class, technical methods, language and so on."[123] Producing a complete "repertoire" of floor plans based on this representation was a matter of combinatorics, finding all planar graphs for a given number of rooms and populating them with labels.[124]

Crucially, graphs justified the necessity for the Ville Spatiale megastructure. Friedman argued that a necessary support for the "repertoire," whose politics we will return to, was a "completely non connected graph" containing at least a number of points equal to the number of rooms the repertoire is being calculated for.[125] This set of points represented an *infrastructure* devised to accommodate all possible floor plans.[126] Friedman examined two architectural possibilities as materializations of this system, which he referred to as a "troglodyte" and a "skeleton" infrastructure. He drew the "troglodyte" infrastructure as a repetitive tiling of orthogonal enclosures that contained points not linked to each other.[127] The "skeleton" model was the inverse: it consisted of all points connected to each other.[128] In the first, architectural configurations emerged by creating links between the enclosures (points); in the second, by cutting links between enclosures (points) through partitions. The choice between the two was to be made based on considerations of the local industrial context. For instance, Friedman wrote, the "troglodyte" model was akin to building processes where "enclosures are built first and then doors and windows are 'punched out.'"[129] Friedman reported on having used this model "to solve construction problems for low cost housing in an African nation" and lauded its benefits for helping him combine mass production with local techniques and the "very rich individual characteristics of a sophisticated culture."[130] The framework model was the Ville Spatiale. It was a strongly linked graph that could accommodate all subgraphs.

Friedman's practice of rarely citing sources starkly contrasts the vast network of architects, artists, and scientists from which his theories were finding inspiration. Thousands of pages of saved typed and handwritten

a

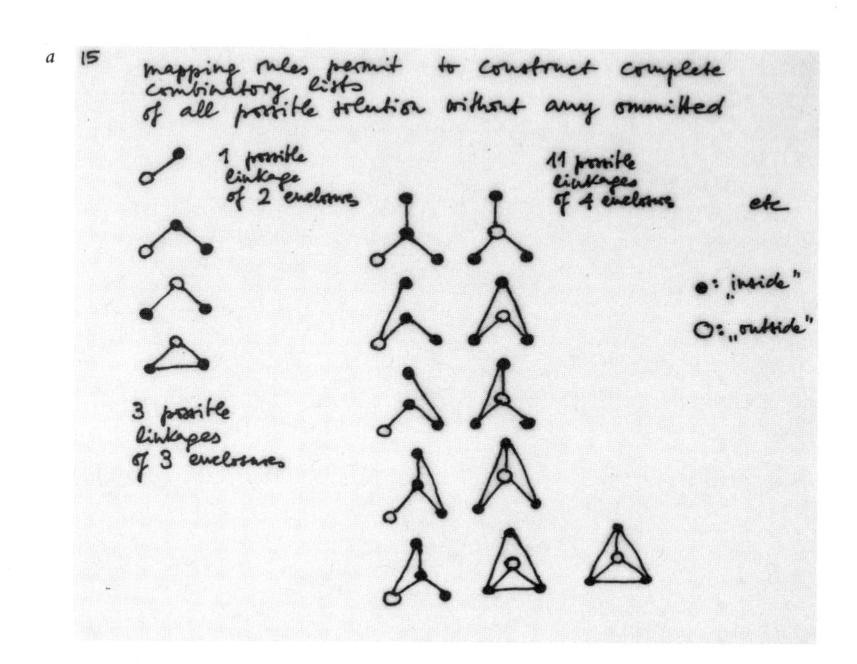

b

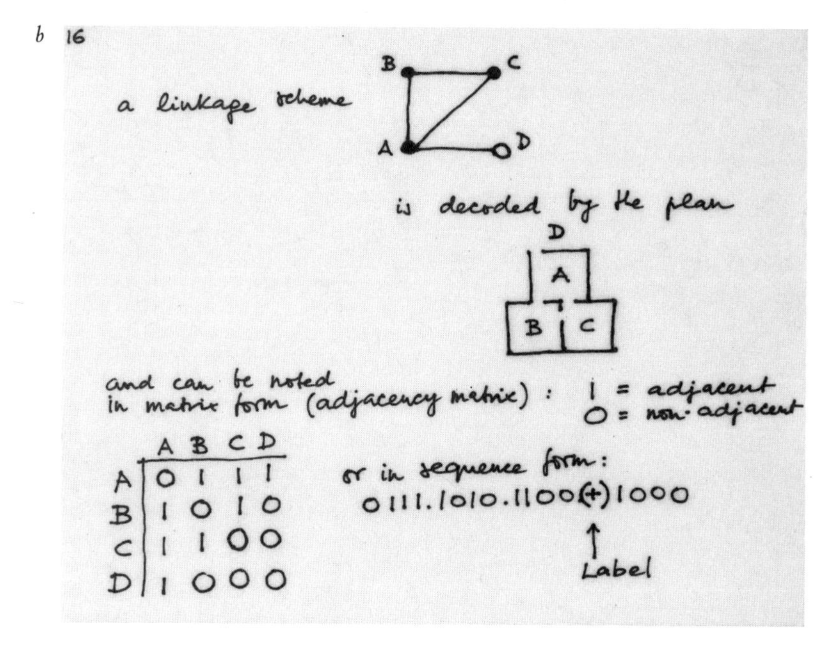

Figure 2.23
"The future user: manual on architecture" (n.d., circa late 1960) by Yona Friedman. Use of graphs to identify the number of statements permissible in an axiomatic theory of architecture (*a*), a schematic architectural plan mapped one-to-one to a labeled planar graph, and then translated to an adjacency matrix and a sequence of numbers (*b*), the infrastructure as a graph and a matrix (*c*), and the "space frame" model (*d*). *Source*: Fond Denise et Yona Friedman. Folder 293, pages 228 (*a*), 229 (*b*), 236 (*c*), 238 (*d*). Courtesy Fond Denise et Yona Friedman, Archives photographiques Jean-Baptiste Decavèle, Marianne Friedman Polonsky, https://www.yonafriedman.org/.

c

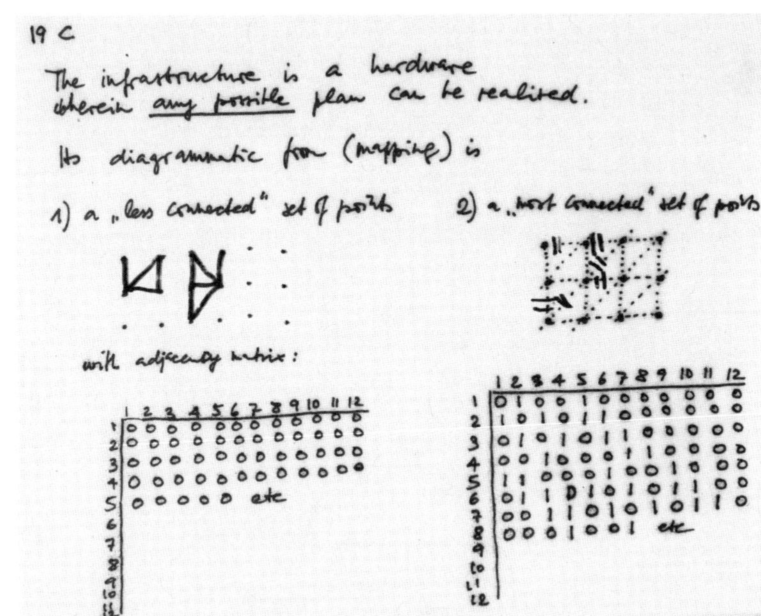

19 C
The infrastructure is a hardware
wherein any possible plan can be realised.
Its diagrammatic form (mapping) is
1) a „less connected" set of points
2) a „most connected" set of points
with adjacency matrix:
1 2 3 4 5 6 7 8 9 10 11 12
1 0 0 0 0 0 0 0 0 0 0 0 0
2 0 0 0 0 0 0 0 0 0 0 0 0
3 0 0 0 0 0 0 0 0 0 0 0 0
4 0 0 0 0 0 0 0 0 0 0 0 0
5 0 0 0 0 0 etc
6
7
8
9
10
11
1 2 3 4 5 6 7 8 9 10 11 12
1 0 1 0 0 1 0 0 0 0 0 0 0
2 1 0 1 0 1 1 0 0 0 0 0 0
3 0 1 0 1 0 1 1 0 0 0 0 0
4 0 0 1 0 0 0 1 1 0 0 0 0
5 1 1 0 0 0 1 0 0 1 0 0 0
6 0 1 1 0 1 0 1 0 1 1 0 0
7 0 0 1 1 0 1 0 1 0 1 1 0
8 0 0 0 1 0 0 1 etc
9
10
11
12

d

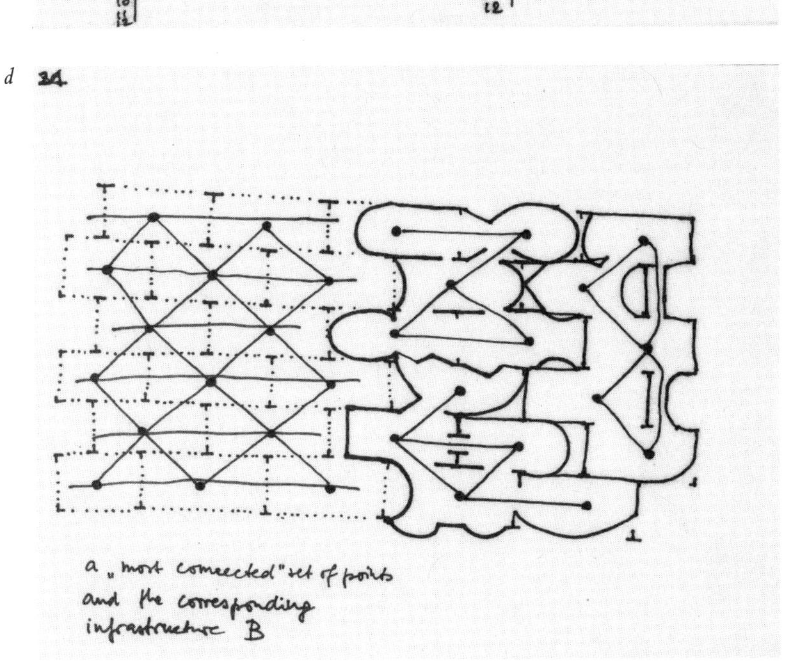

a „most connected" set of points
and the corresponding
infrastructure B

letters and notes in Friedman's personal archive reveal that he was consciously and steadily building a vast international network and was eager to be part of the latest conversations in art, science, and technology. This archive reveals that Friedman was introduced to the graph from two of its most active proselytizers. When he was teaching at the University of Michigan, in Ann Arbor, Friedman developed a close friendship with Frank Harary, or as some liked to call him, "Mr. Graph Theory." Harary and Friedman shared a common friend, artist and mathematics enthusiast Anthony Hill, who in 1963 had written with Harary a paper summarizing their work on the still intractable problem of crossings in a complete graph—the so-called "brick factory problem."[131]

Hill was a member of the British Constructionist Group together with Victor Pasmore, Kenneth Martin and Mary Martin, Stephen Gilbert, and others. Friedman's archive contains dozens of letters with Hill in 1964 and 1967, one year before Hill published the famous anthology *DATA* (standing for *Directions in Art, Theory, and Aesthetics*).[132] In the tradition of the avant-garde publication *Circle*, published by Faber and Faber in 1937, *DATA* featured a collection of polemical essays from plastic artists who explored some intersection of related themes: "constructive, concrete, kinetic, structurist and synthetist."[133] These were interspersed with contributions from "specialists" from philosophy, mathematics, physics, engineering, sociology, and urbanism.[134]

Friedman contributed an essay titled "Reflections on the Architecture of the Future: Criteria for Town-Planning," in which he outlined the fundamental principles of his theories for spatial mobility and infrastructural urbanism, with little mathematical exposition.[135] Hill's essay on the other hand was rife with mathematical ideas about topological symmetry and aesthetic measures, including a famous use of graphs to analyze Piet Mondrian's paintings. Hill's interest was in what he referred to as "paragrams": diagrammatic inscriptions that accompanied the presentation of ideas in linear prose. Paragrams, Hill suggested, were a recurrent but underdiscussed device in modern art; from Alfred Barr's family trees to Picabia's playful charts to Iannis Xenakis's organigrams, to mention some of the examples Hill enumerated. But really the discussion of "paragrams" was a strategic ploy to get to his favorite topic: graphs. Hill constructed a lineage of the modern paragram that traversed Leibniz's isomorphisms between geometry and logic, Euler's laying of the foundations of graph theory, and Charles Sanders Peirce's discussions of graphs as diagrammatic forms. This history elevated a familiar suspect. Hill wrote:

> Certainly topology is found to be a constant link in all these areas, but the specific context for the paragram is "graph theory," which is concerned among other things with the study of line and point

structures. . . . In an increasing number of fields graphs are becoming a major tool and might be described as a new lingua franca.[136]

The rest of the essay was concerned with a revision of American mathematician George David Birkhoff's formulas for aesthetic measure, which was in good currency in contemporaneous debates around "information aesthetics" advanced by influential philosophers such as Abraham Moles and Max Bense.[137] Unlike Birkhoff, whose measures concerned the visual presentation of shapes, Hill was interested in *topological* symmetry and asymmetry: with symmetries that had to do with the ability or inability to distinguish points within a graph. Hill claimed: "Whereas metrical symmetry as a quality is in some measure always relative to an absolute (the ideal), involving deviations, approximations etc., topological symmetry, is at all times an absolute and invariant quality."[138]

Although Hill's and Friedman's contributions to *DATA* seem far apart in style of argumentation and mathematical fluency, the two shared a long and intimate connection through graphs. A folder of personal correspondence from 1964 reveals lengthy and enthusiastic letters enclosing pages on pages of sketches, in which Hill and Friedman think through topological formulations of symmetry and packings of polyhedra. Friedman sends "conjectures" that Hill responds to with patience and encouragement, correcting terminology and mathematical errors, and drawing graphs over graphs.[139] One letter also includes a request from Friedman for ten dozen "Bombay spangles" from the India Craft store at Oxford Street near the corner of Tottenham Court Road in London. Friedman promises to send back "theorems (?)" in the form of model photography of these chains.[140] In October 1964, Hill's letters on topological symmetry accompanied Friedman on his flight to the United States, the airplane being an ideal setting for him to think about the mathematics with a "clear head."[141] That month Hill also reports to Friedman having shared ideas on topological symmetry with members of our Blanche: C. A. B. Smith and Bill Tutte.[142]

The thread continues in the archive in 1967, with Hill sending Friedman an update on the production of *DATA*. Hill reports "straining [his] tiny brain cells on graph theory" and regularly meeting Frank Harary and "his boys" at University College London before Harary's departure for the University of Michigan. Hill also mentions Harary having inquired at the University of Michigan Press about the publication of the *Architecture Mobile* but with no success. Pergamon Press, the publisher of Frank Malina's then new journal *Leonardo*, for which Hill became a coeditor, would be an alternative.[143]

A few days later, Friedman would write back to Hill that he was planning on turning down a full professorship with tenure that Michigan offered him: "I am very proud of this, but I think not to accept it, as I don't feel myself old enough to retire to a small town, and would like to stay in Paris till might [sic] last possibility not to be a victim of hunger. But I think, I will accept something on a partial basis, for four weeks, for example."[144]

In fall 1967, Friedman would perform these seminars in crutches after an accident arriving to New York City led him to the hospital at Ann Arbor and staying with the Hararys for healing and mathematical rumination. The visiting appointments continued in January and February 1968. Throughout that time, Hill remained an enthusiastic supporter of Friedman's explorations, proofreading his work with ambivalence about translating his "inter-English" to "national-English" and checking the mathematics.[145]

The correspondence takes us to 1968 when Friedman published a "condensed version" in the architecture magazine *ARCH+* of the seminars he had offered between 1964 and 1967 in institutions such as the University of Michigan, Harvard University, and the Université de Montréal.[146] *Pour une architecture scientifique* was the culmination of Friedman's exchange with international networks, key nodes of which had been Hill and Harary. In his lecture notes that formed the corpus of the book, Friedman advocated for a wholesale reform of architectural education. He polemically presented an "ordering process" aiming to transform architecture from a "form of witch-doctorship, a set of uncoordinated kitchen recipe-type knowledge into a well-ordered discipline."[147] Delegating the material aspects of architecture to the artisan or the manufacturer, Friedman defined the disciplinary task of the architect as one of "assembling catalog elements" (industrialized architectural components, prefabricated houses, and so on) to produce spatial configurations.[148] Invoking an unreferenced quote by Walter Gropius, who allegedly characterized architecture as "sociology's hardware," Friedman described a new way of representing architecture:[149] "A diagram of points and links (I will call it a network instead of its correct mathematical name: a graph), can give a simplified but understandable image of any architectural plan."[150]

Graphs would not only order architecture but also render it teachable. The "teachability" of his theory was a central concern for Friedman, who dedicated two sections of *Pour une architecture scientifique*'s first chapter to developing a counterproposal to the "apprenticeship" model of instruction where the student learns by imitating a "master."[151] Friedman found this model precarious because of its dependence on the personality and preferences of the master. Transforming architecture from a "prenticeable" to a "teachable" discipline meant addressing its lack of clear rules and its reliance on "tricks of the trade."[152] Architecture as a "teachable science" would contain "strict rules" that could be transmitted to students in the same way, despite differences in the personalities of instructors, and could be written down in a textbook. Friedman invoked the example of arithmetic.[153]

a

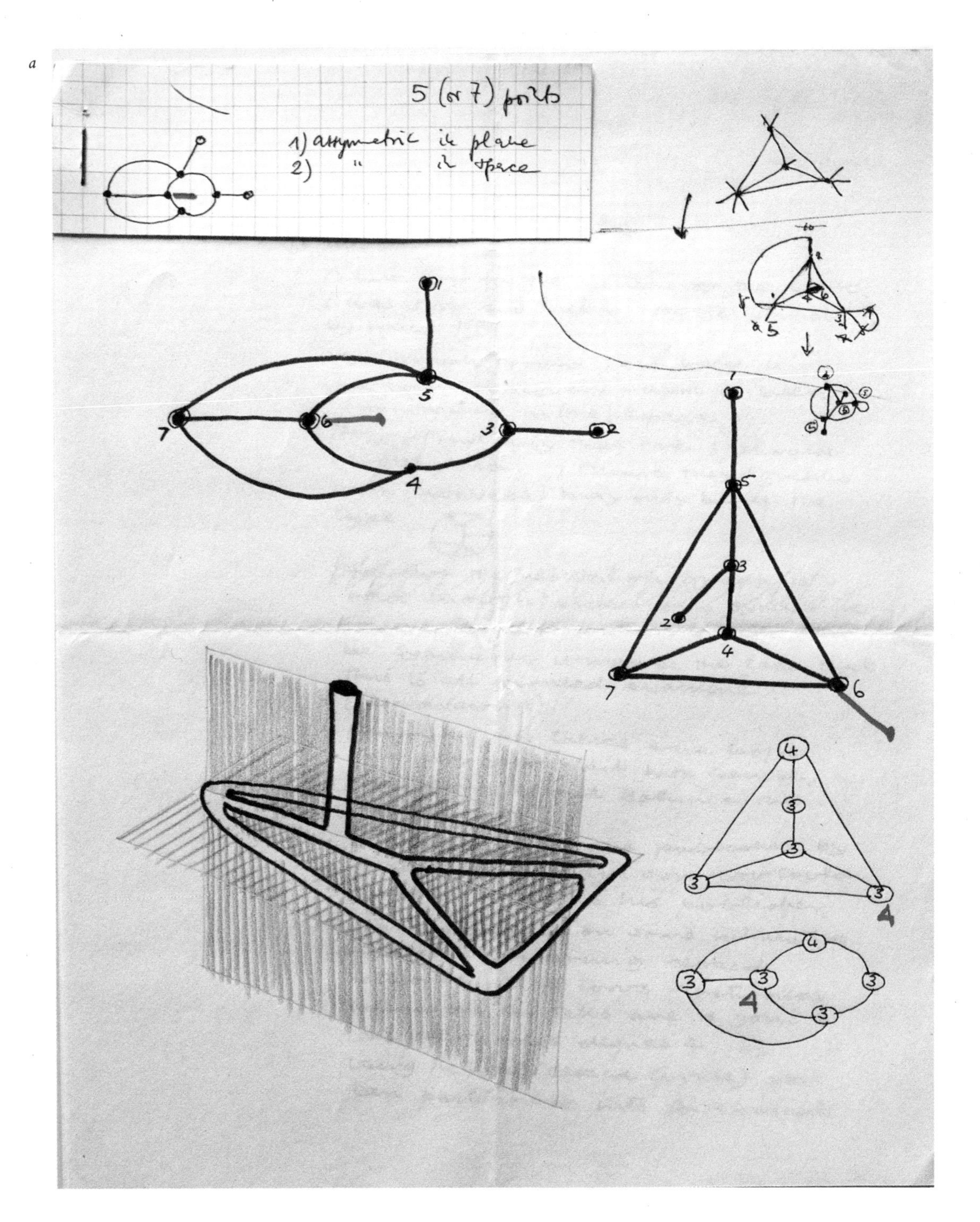

Figure 2.24 (*above and following pages*)
Sketches from Friedman's letter to Hill on 5 September 1964, illustrating a "theorem" Friedman proposed for finding a minimal asymmetric structure. He suggested that this structure was the simplex *n*+1-hedron in which every node ("sommet") had a different degree ("affix") (*a*). A few days later, Hill sketched back a refutation, and a problem about the smallest number of additions to a structure to produce asymmetry (*b*, *c*). *Source*: Fond Denise et Yona Friedman. Folder 283, pages 247 (*a*), 229 and 233 (*b*, *c*). Courtesy Fond Denise et Yona Friedman, Archives photographiques Jean-Baptiste Decavèle, Marianne Friedman Polonsky, https://www.yonafriedman.org/.

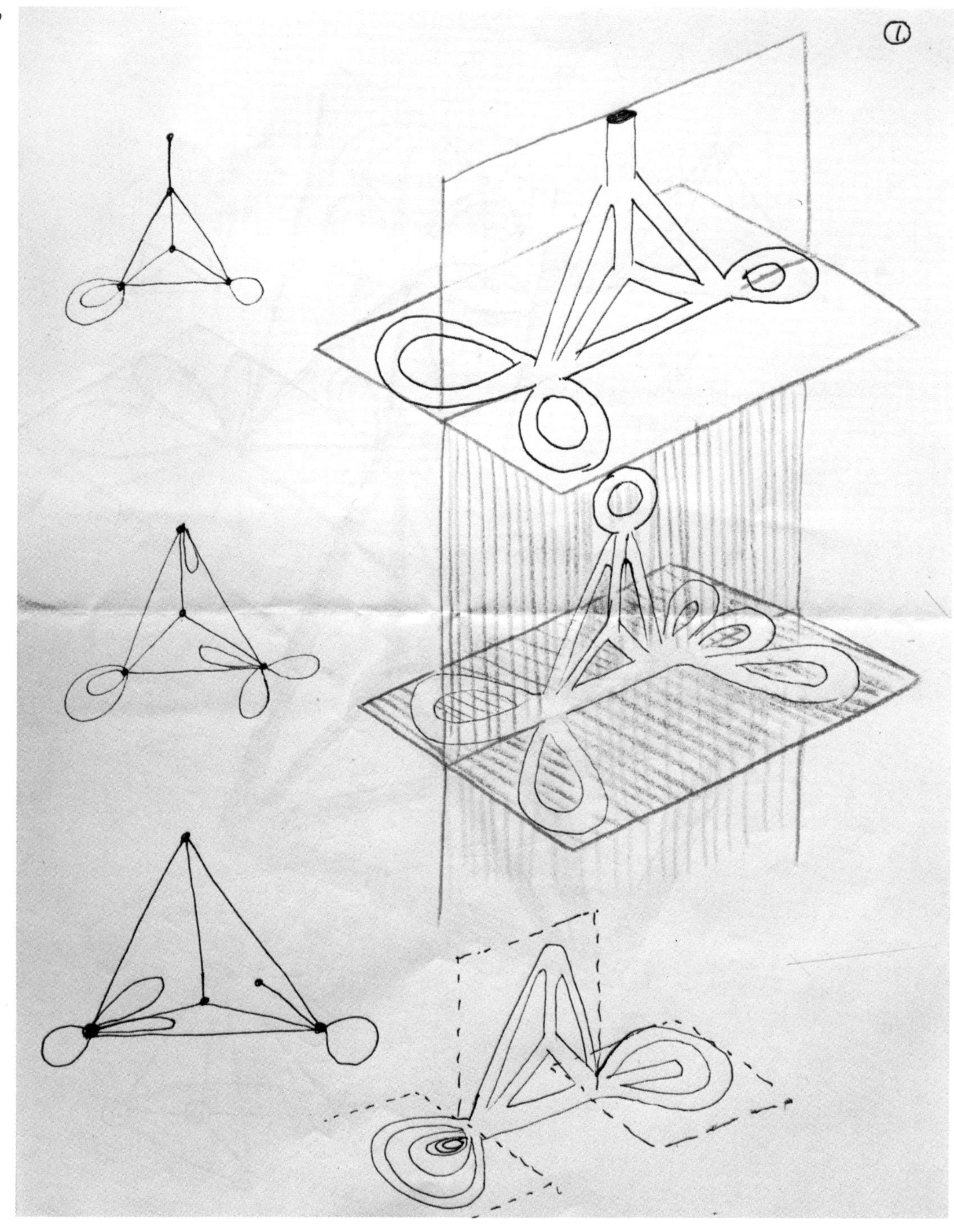

(4)

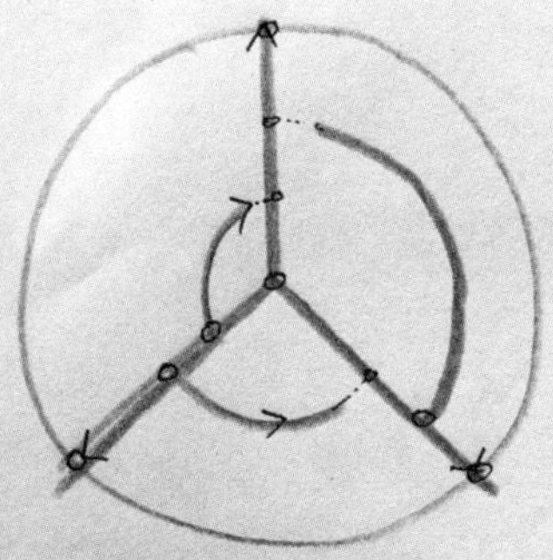

so that the minimal figure that now suggests itself is:

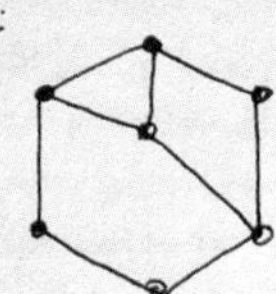

which since we do not allow nodes of degree 2 will have to be

B

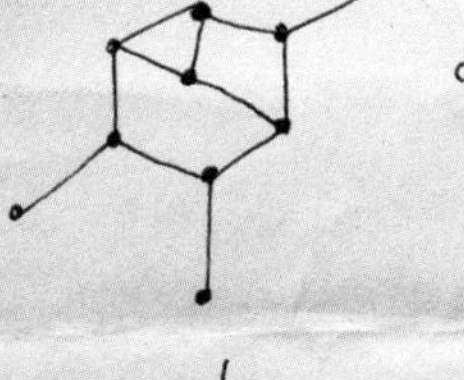

or

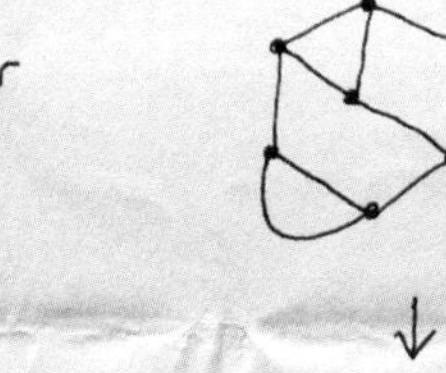

C

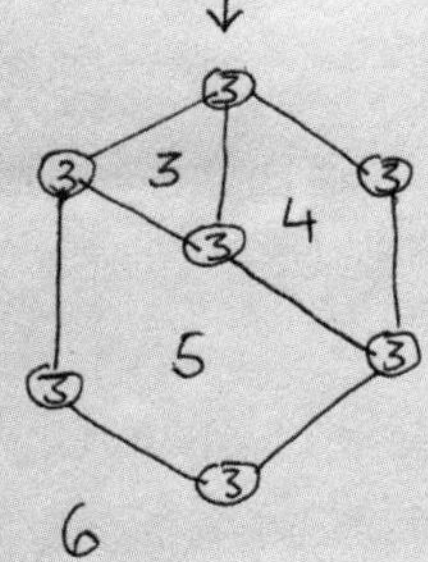

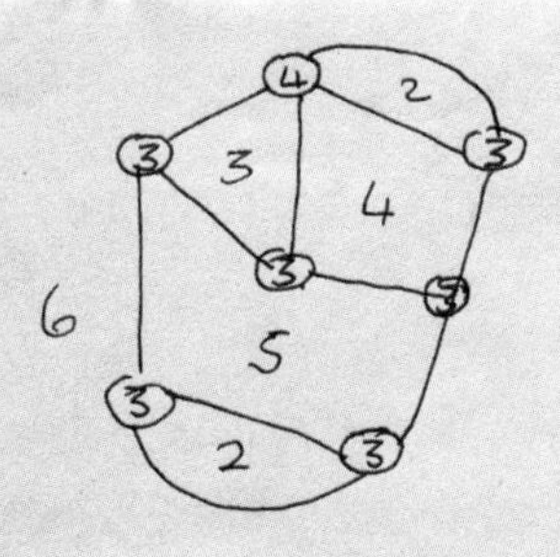

this would appear to be the answer ?
for three different types of graph
A = polyhedral
B = with end points
C = a map or polyhedron allowing 2 gons.

[illegible] 8.9.64.

a

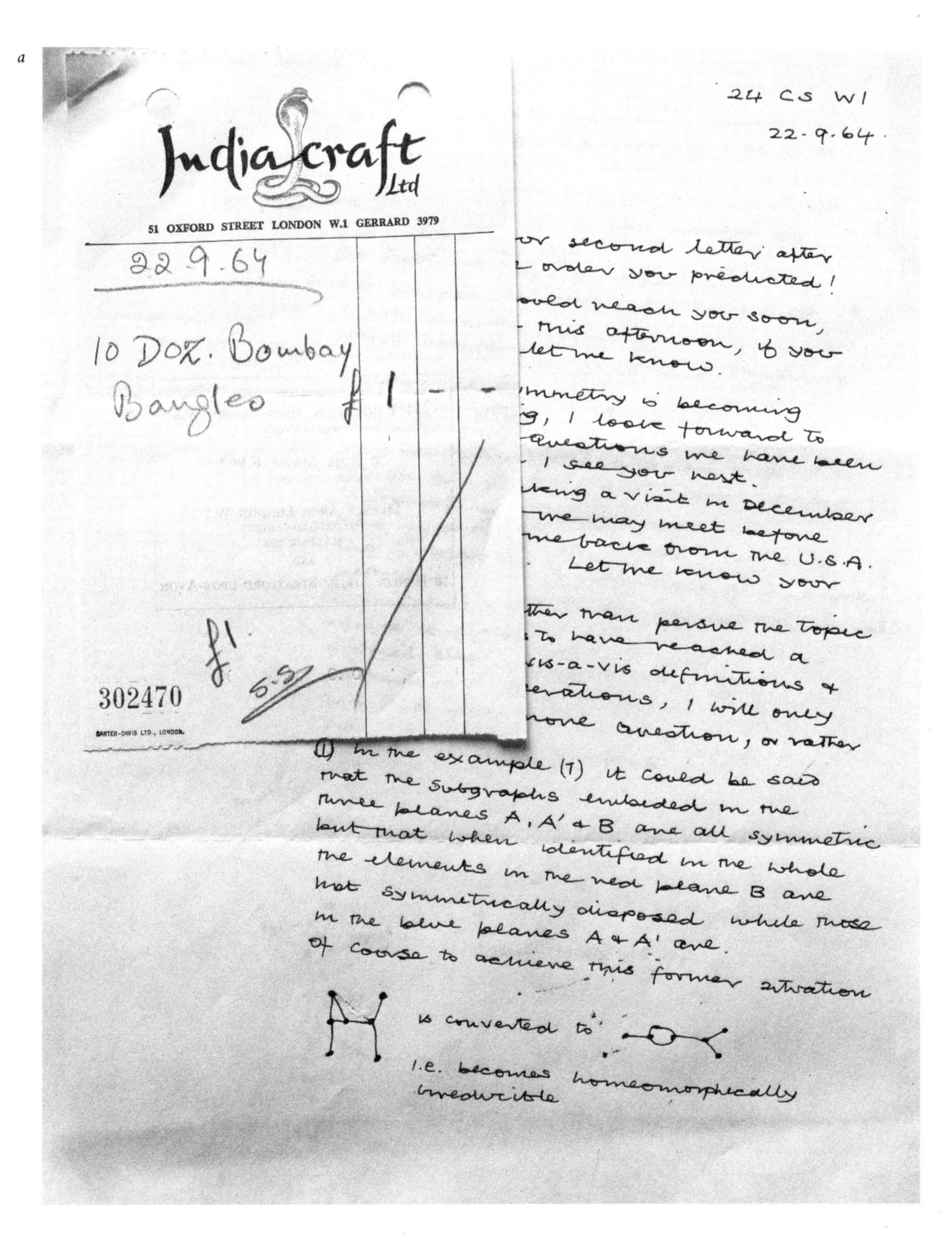

Indiacraft Ltd

51 OXFORD STREET LONDON W.1 GERRARD 3979

22.9.64

10 DOZ. Bombay Bangles £1 – –

£1

5.S

302470

BARTER-DAVIS LTD., LONDON.

24 CS W1

22.9.64.

...or second letter after
...order you predicted!
...ould reach you soon,
... this afternoon, if you
...let me know.

...mmetry is becoming
...g, I look forward to
...questions we have been
... I see you next.
...king a visit in December
... we may meet before
...me back from the U.S.A.
... Let me know your

...ther than persue the topic
... to have reached a
...vis-a-vis definitions &
...erations, I will only
...none question, or rather

(1) In the example (7) it could be said that the subgraphs embeded in the three planes A, A' & B are all symmetric but that when identified in the whole the elements in the red plane B are not symmetrically disposed while those in the blue planes A & A' are.
Of course to achieve this former situation

is converted to

i.e. becomes homeomorphically irreducible

Figure 2.25
The correspondence on topological symmetry continues, with a receipt of a 10 dozen Bombay bangles attached (*a*). Hill's sketches of a conjecture on "asymmetrical finite homeomorphically metrizable structures" in a letter to Friedman on 25 September 1964; Hill wrote that "to such a polyhedron one could attach end points, trees, loops, etc. and presumably the structure would continue to be asymmetric" (*b*). *Source*: Fond Denise et Yona Friedman. Folder 283, pages 202 (*a*), 196 (*b*). Courtesy Fond Denise et Yona Friedman, Archives photographiques Jean-Baptiste Decavèle, Marianne Friedman Polonsky, https://www.yonafriedman.org/.

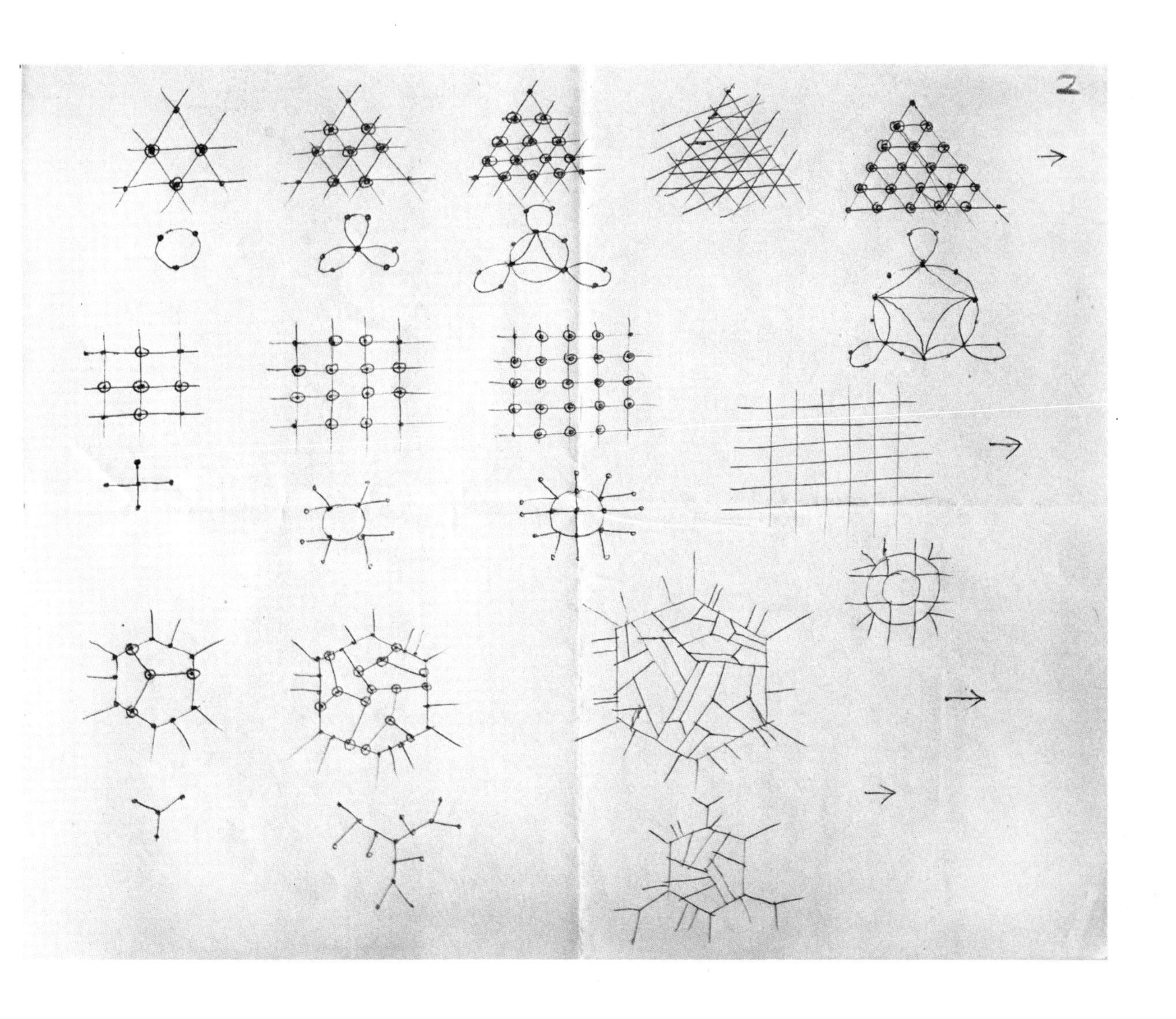
2

a

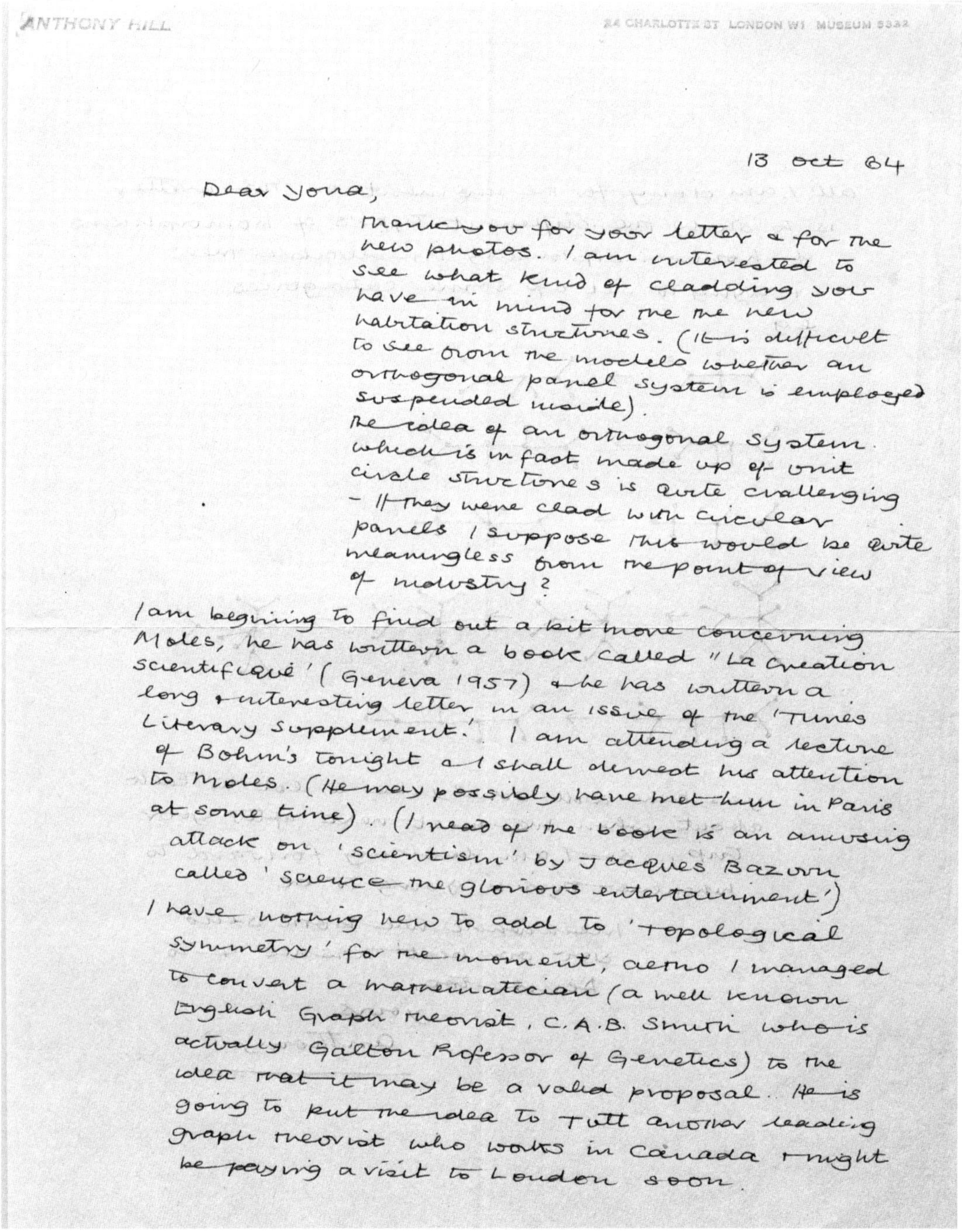

ANTHONY HILL 24 CHARLOTTE ST LONDON W1 MUSEUM 3332

13 oct 64

Dear Yona,

Thank you for your letter & for the new photos. I am interested to see what kind of cladding you have in mind for the the new habitation structures. (It is difficult to see from the models whether an orthogonal panel system is employed suspended inside).

The idea of an orthogonal system which is in fact made up of unit circle structures is quite challenging – If they were clad with circular panels I suppose this would be quite meaningless from the point of view of industry?

I am begining to find out a bit more concerning Moles, he has written a book called "La Creation Scientifiqué" (Geneva 1957) & he has written a long & interesting letter in an issue of the 'Times Literary Supplement.' I am attending a lecture of Bohm's tonight & I shall direct his attention to Moles. (He may possibly have met him in Paris at some time). (I read of the book is an amusing attack on 'scientism' by Jacques Bazun called 'Science the glorious entertainment')

I have nothing new to add to 'topological symmetry' for the moment, altho I managed to convert a mathematician (a well known English Graph theorist, C.A.B. Smith who is actually Galton Professor of Genetics) to the idea that it may be a valid proposal. He is going to put the idea to Tutt another leading graph theorist who works in Canada & might be paying a visit to London soon.

Figure 2.26
Hill's letter to Friedman on 13 October 1964 on his progress with topological symmetry and his interactions with mathematicians C. A. B. Smith and William Tutte. *Source*: Fond Denise et Yona Friedman. Folder 283, pages 161 (*a*), 162 (*b*). Courtesy Fond Denise et Yona Friedman, Archives photographiques Jean-Baptiste Decavèle, Marianne Friedman Polonsky, https://www.yonafriedman.org/. All rights reserved.

b

all I am doing for the moment on the matter is to study the different types of isomorphisms of homeomorphically irreducible trees & trying to set up some categories.

e.g.

yes there will surely be a lot to talk about, when we meet next after your trip, so I am looking forward to whenever this is going to be.

Meanwhile we both send you our best wishes & to Denise too,

yours

anthony

a

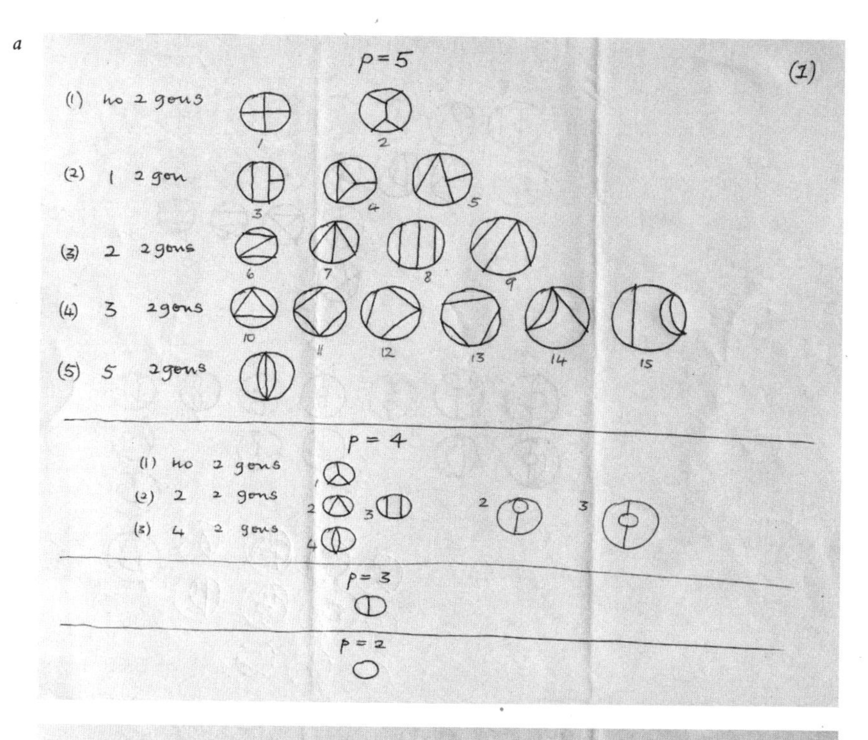

b

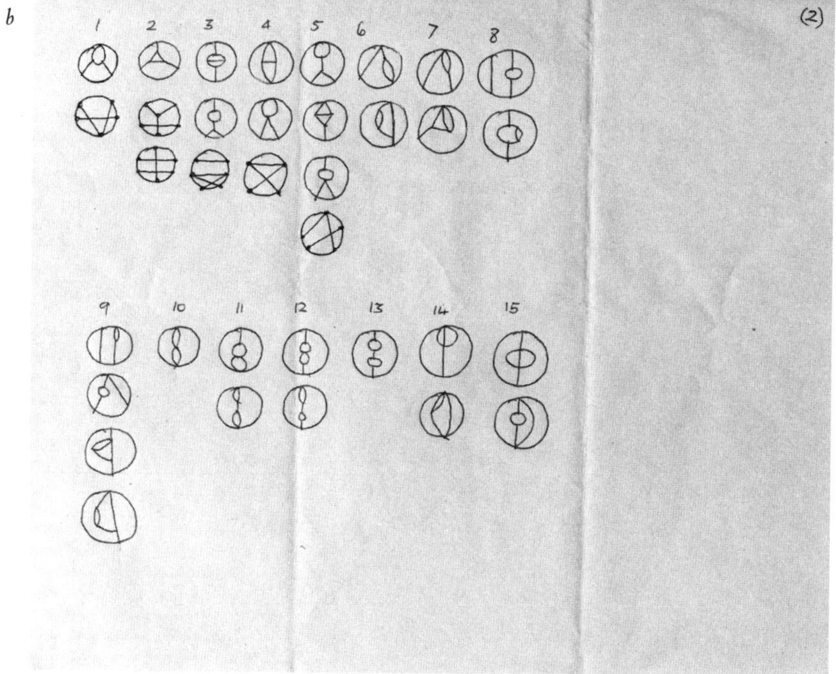

Figure 2.27
Hill's letter to Friedman on 1 October 1965 illustrating his progress on "orthogonal structures with zero information content" and "on the problem of enumerating all polygonal networks." *Source*: Fond Denise et Yona Friedman. Folder 283, pages 516 (*a*), 517 (*b*). Courtesy Fond Denise et Yona Friedman, Archives photographiques Jean-Baptiste Decavèle, Marianne Friedman Polonsky, https://www.yonafriedman.org/. All rights reserved.

Architecture, Friedman continued, in a statement that it is tempting to connect with the discussion on school education earlier, could be taught at the primary level. Everyone would learn how to *read* architecture at school, and those wishing to become professional architects would go to university to learn more about how to materialize it by turning diagrams of spatial structure, graphs, into buildings. "There is nothing unusual about this situation," he wrote, "mathematicians and laymen use the same methods and the same rules, but on different levels. Anyone can read a weather map, but only a scientist can draw one."[154]

In his preface to *Pour une architecture scientifique*, philosopher of art and director of Belfond's Art-Action-Architecture series Philippe Sers chastised French architectural audiences for remaining oblivious to Friedman's ideas while he was disseminating them in the United States, Russia, Japan, and Cuba. Friedman, he caustically remarked, seemed to have achieved his childhood wish to "be a stranger."[157] Yet despite his tirelessly outward-facing activity, Friedman's sober sketching of point-line skeletons in the forms of graphs continued to participate in a vibrant French cultural scene concerned with space frames and the architectural protocols that governed them—or, to use French journalist and critic Michel Ragon's term, with "spatial urbanism."[158] These architectural experiments and the interdisciplinary groups that pursued them and of which Friedman was part—for example, the GEAM or the Groupe International d'Architecture Prospective (GIAP) founded by Ragon—promoted thinking about architecture in terms of the combinatorics of discrete spatial units. This is perhaps most clearly articulated in the article "Une théorie pour l'occupation de l'espace" published in 1962 in the French journal *Architecture d'aujourd'hui* by GEAM member and Friedman's close collaborator Eckhard Schulze-Fielitz. In the article, Schulze-Fielitz famously proposed to view space as a "macro-material" which can be reduced to a "few elementary particles," and argued that the characteristics of the material could be determined by the "combinatorial possibilities" of these spatial atoms.[159]

A keen observer of new technological and scientific currents, Friedman brought together the architectural expression of space frames with a mathematical formulation that vowed to clarify their principles. Space frame explorations were crucial in supporting Friedman's claims that graphs represented architecture in a non-distortive manner. The graph's discrete points matched the discrete units in which space frames segmented three-dimensional space. Its lines represented the act of combining and configuring these units, of doing architecture in a space frame world. Friedman's seemingly arbitrary architectural axiomatics succinctly described the tenets of a prospective form of architectural practice that was brewing among space frame experimentalists.

But it was also about the graph's looks. Graphs' points and lines visually resembled the Ville Spatiale's rods and nodes. In *Pour une*

a

CHAPITRE V.3.
Diagrammes de croissance

Les schémas possibles de croissance (partitions) sont représentés par:

A: le schéma (en n étapes)

B: le G_n graph correspondant

C: le $G_{n,1}$ graph correspondant

D: le code résultant des permutations des degrés du graph $G_{n,1}$

A	B	C	n	D
			1	11
			2	222
			2	121
			3	2323
			3	3333
			3	2231
			3	1221
			3	1322
			3	2222
			3	1311

4

C	D	C	D	C	D
	22224		23334 33334		12223 12223 12223
	11224 11224		23333 23333		
	11114		22332 22332		
	13231 13231		23232 23232		
	11123		23324 23324 23324		
	43434 43434		23223 23223 23223		
	24334 24334		22222		
	13234 13334		12221 12221		

Figure 2.28
Notes for the seminars by Yona Friedman at University of Michigan, Harvard University, and the Université de Montréal (1967). Studies of architectural layouts using permutations of a planar graph (*a*) and network studies (*b*). *Source*: Fond Denise et Yona Friedman. Folder 264, pages 29 (*a*), 23 (*b*). Courtesy Fond Denise et Yona Friedman, Archives photographiques Jean-Baptiste Decavèle, Marianne Friedman Polonsky, https://www.yonafriedman.org/.

b

16-4-68

CHAPITRE III.

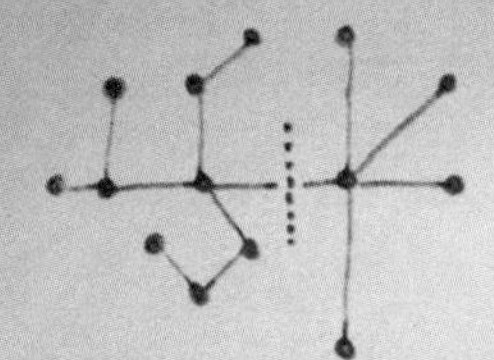

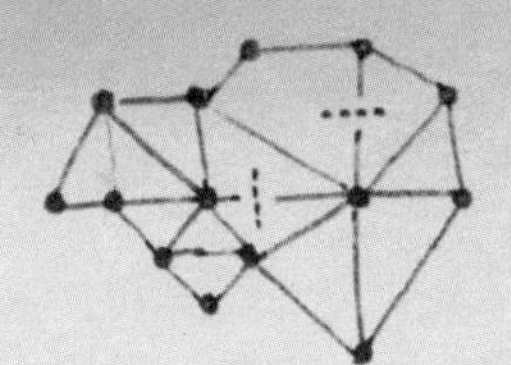

Arbre: une seule coupure partage le système en deux parties indépendantes

Réseau: plusieurs coupures sont nécessaires pour partager le système en deux parties indépendantes

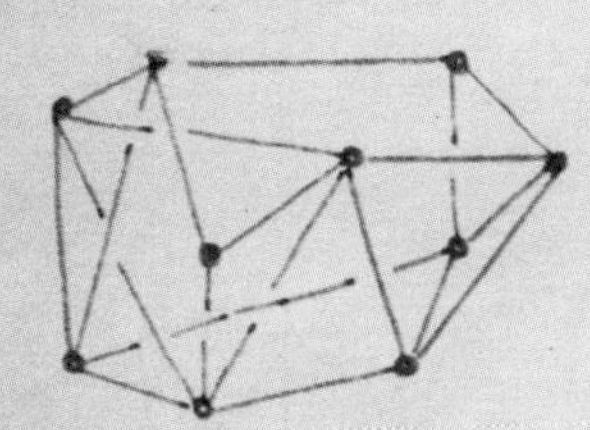

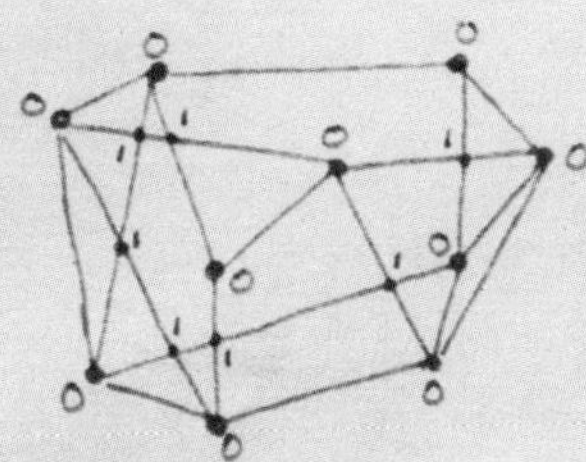

Un réseau planaire peut être transformé en un réseau spatial (non-planaire) si l'on considère une partie des vertices comme des „croisements" et l'autre partie comme des „passages à niveaux".

Nous pouvons donc étiqueter, à titre d'exemple, les croisements par (0) et les passages à niveaux par (1).
Tout réseau planaire peut être transformé en réseau spatial par l'étiquettage (labeling) approprié.

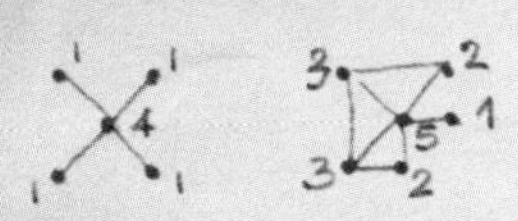

le degré d'un point est indiqué par le nombre des autres points liés au point en question.

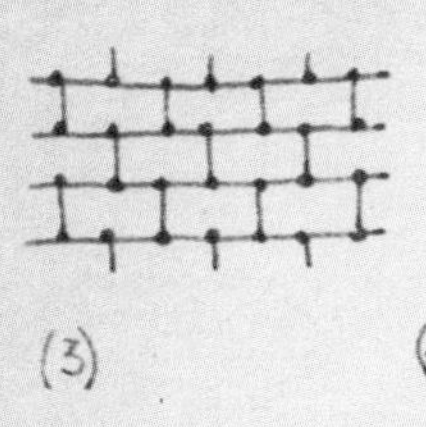

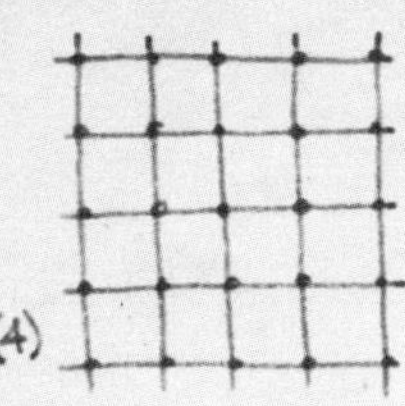

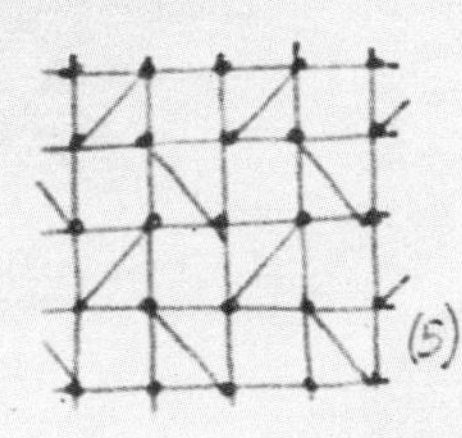

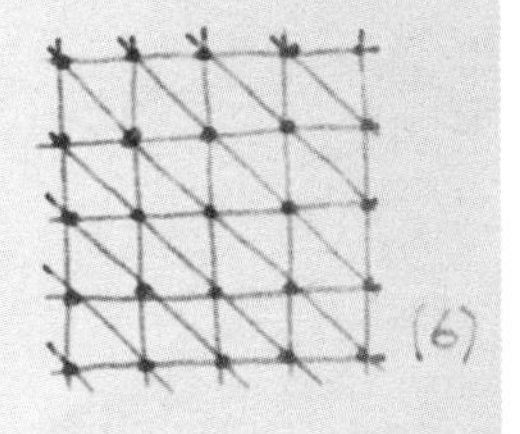

Réseaux planaires homogènes de degré (3), (4), (5) et (6).

Interlude 3

There was something so attractive, so intuitive to graphs, Friedman remembers, that made them stand apart from other math. "Graph theory," Friedman once told me recounting his meetings with Harary, "for certain Mathematicians is not Mathematics, it is not Logic, but it is a very useful tool. It is useful for people to experiment."

That was Harary's project. He was unlike some of his predecessors, like the author of the first graph theory book Dénes Kőnig, who were nervous about the graph's "geometric" presentation and reassured the readers that graphs were not visual but abstract—they could be thought of in terms of sets or matrices.[155] Harary treated graphs as aesthetic objects.

While teaching at the University of Waterloo in the fall of 1970, Harary replicated an experiment by Claude Facheux at the University of Paris, on the psychological aspects of pattern recognition.[156] The experiment involved asking students to draw all tree graphs corresponding to 8 points, revealing the number of options to the first group and concealing it from the second. The goal was to see which graphs were bound to be omitted and correlate the omission with mathematical properties such as symmetry. Upon replicating the experiment, Harary added a step where the students were asked to rank the different tree graphs by aesthetic preference and assign to them labels such as "prettiest," "ugliest," "funniest," "sexiest."

Graphs engaged the mind as much as they did the eye. They were objects of intellect, and they were objects of aesthetic desire.

Tree	Ranking			
	Prettiest	Ugliest	Funniest	Sexiest
1		12		
2 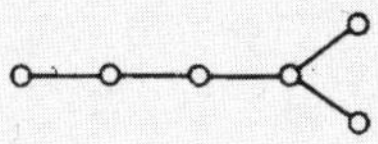	1	1	3	7
3 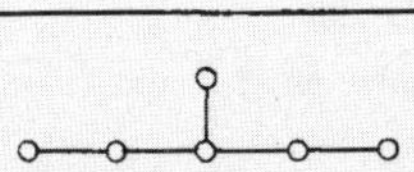		12	7	2
4 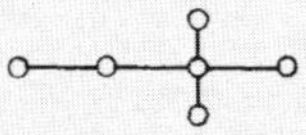	5	7	6	7
5 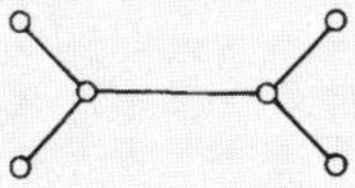	12		10	15
6 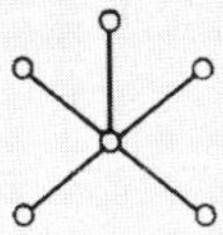	18	4	10	5

Figure 2.29
Source: Frank Harary, "Aesthetic Tree Patterns in Graph Theory," *Leonardo* 4, no. 3 (1971): 229.

Figure 2.30
Notes and sketches on graphs (n.d.) *Source*: Fond Denise et Yona Friedman, Folder 296, page 337. Courtesy Fond Denise et Yona Friedman, Archives photographiques Jean-Baptiste Decavèle, Marianne Friedman Polonsky, https://www.yonafriedman.org/.

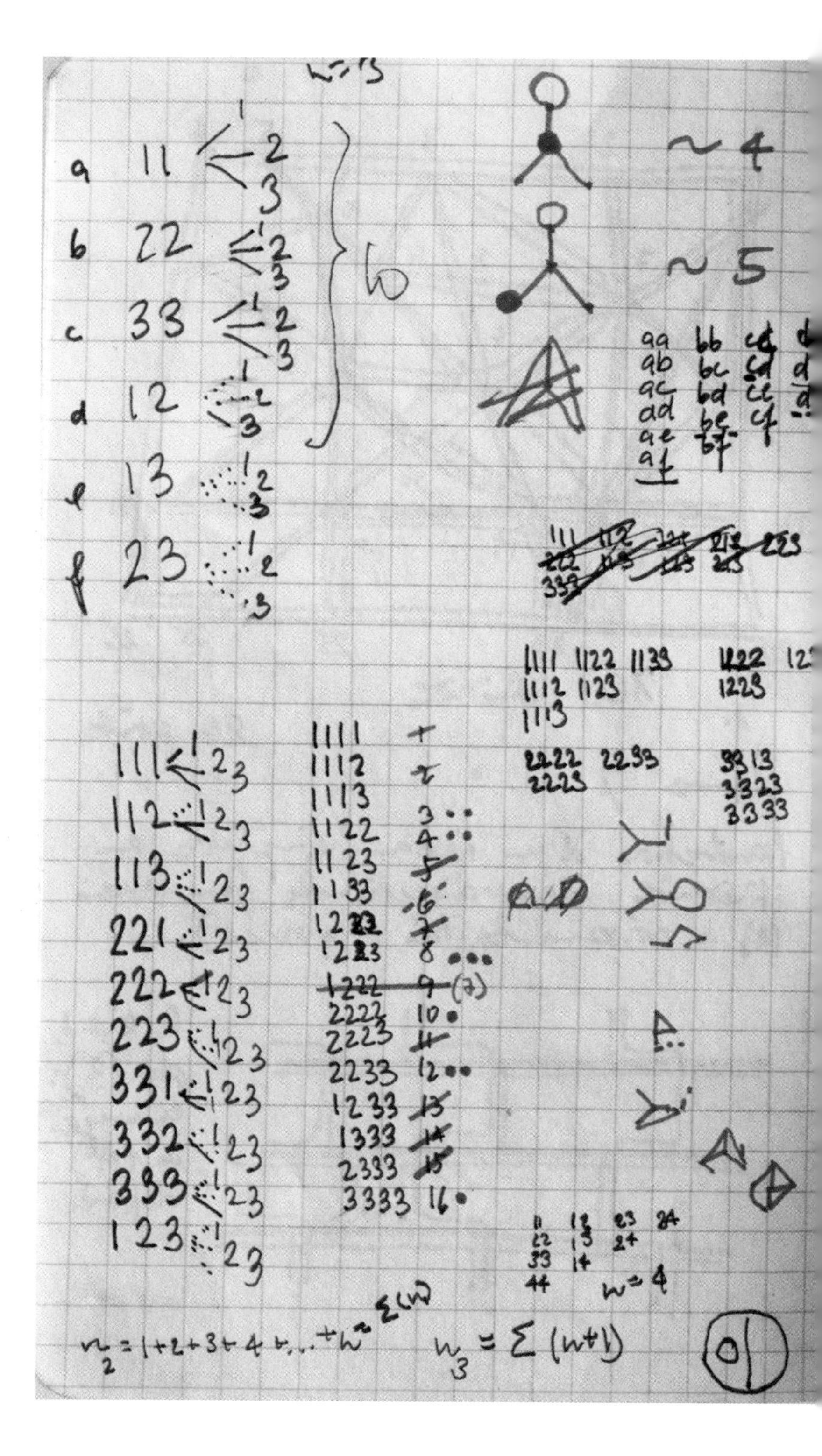

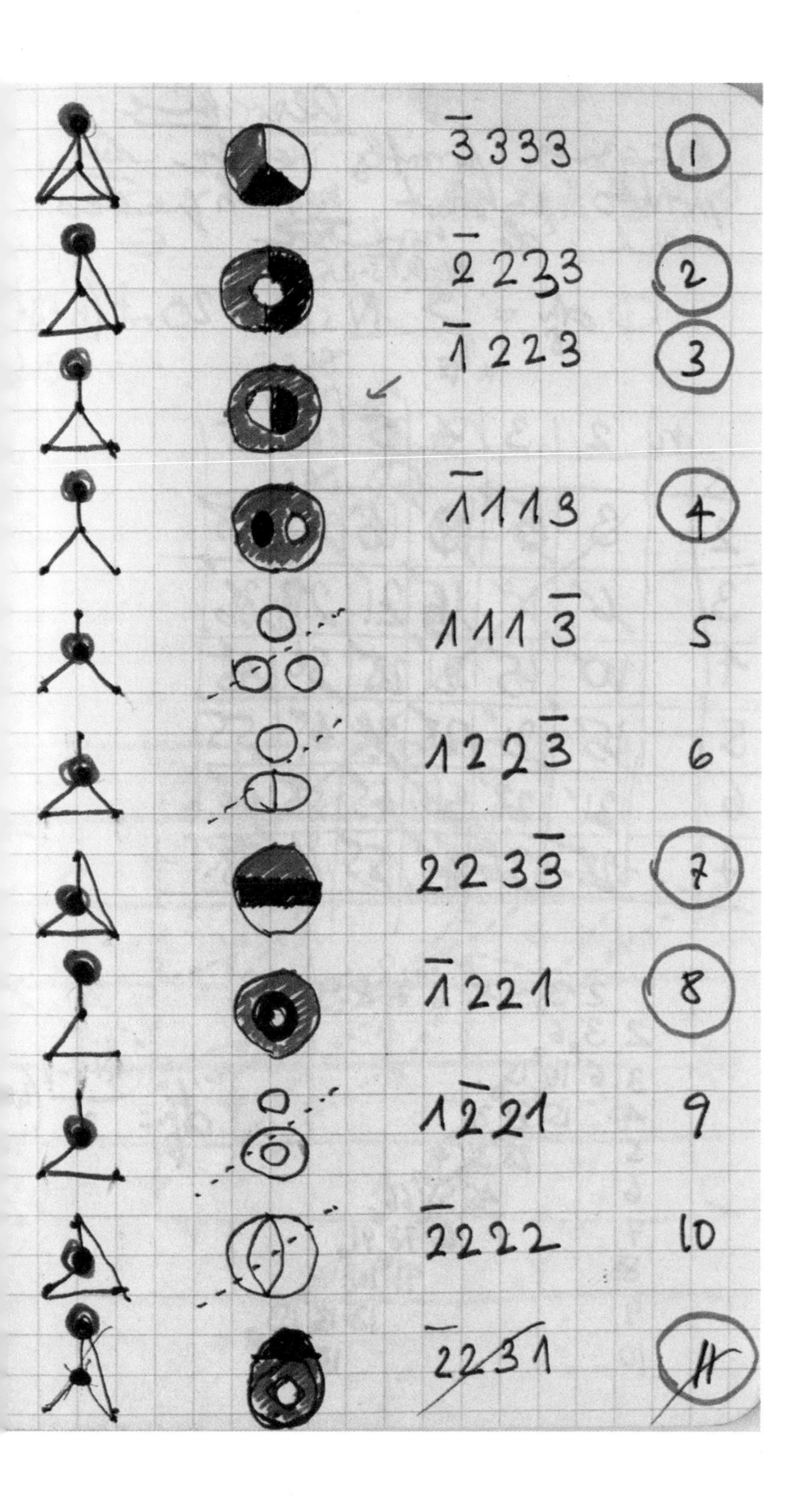
3̄333
1
2̄233
2
1̄223
3
1̄113
4
1113̄
5
1223̄
6
2233̄
7
1̄221
8
12̄21
9
2̄222
10
2̄231
11

architecture scientifique, Friedman drew graphs in the same skeletal visual language in which he sketched his famed architectural megastructures. Graphs and the Ville Spatiale, the architecture and its abstraction, *looked* the same. That the graph was a purely mathematical and not a visual (aesthetic) object was unequivocally conveyed by Friedman's claim that, despite its many illustrations, *Pour une architecture scientifique* was "without pictures."[160] Yet, leafing through dozens of hand-drawn illustrations of points and lines, alongside the occasional abstractions of the Ville Spatiale infrastructure in an identical graphic language, one can see the workings of the same hand and eye. Friedman's flirtation with mathematical abstraction did not interrupt, but instead they enhanced, the development of a signature visual language.[161]

Isomorphisms

The suspicion that Friedman's mathematical theory was precisely what he had made a reputation for critiquing—a *pseudo-theory*—echoed in a report by École Nationale Supérieure des Beaux-Arts student league member Bernard Huet.[162] Huet was among the audience members of the two-day International Dialogue of Experimental Architecture (IDEA) symposium at Folkestone, UK, organized by the Archigram group. Friedman was invited as one of "Europe's creative nuts," as the IDEA poster put it, along with Cedric Price, Ionel Schein, Hans Hollein, and others.[163] In reporting on IDEA for the student magazine *Melp! (Melpomene)*, Huet criticized Friedman's reliance on what he referred to, in his report, as a "pseudo-mathematical vocabulary of a disarming naiveté" used to justify formal ideas that Huet found to be "quite limited."[164] Huet applauded Friedman's effort to ground architecture in mathematics and logic—he mentioned Christopher Alexander as an instigator of such efforts—but dismissed Friedman's mathematics as too obvious an afterthought, a post-rationalization of his aesthetic attraction to tantalizing high-tech space frames that were capturing many of his contemporaries' attention.

He wasn't wrong. Friedman, of course, claimed the opposite—that mathematics gave him a language to articulate the principles of his work that had been there from the outset.[165] But it was graphs' isomorphism with space frames that drove him to them. That—maybe—made graphs appear so irresistibly attractive in that meeting with Harary at Ann Arbor somewhere in the mid-1960s, a meeting that was followed by close friendship with Harary and with graphs. This isomorphism was not about mappings between structures hiding beneath the appearance of things. Or rather, not just about these mappings. It was an isomorphism that, paradoxically, operated on the surface.

The "equality of form" between graphs and the Ville Spatiale was literal. It was about the kind of treacherous visual similitude that pulls things together, that captures the eye before the mind has time to think, that renders abstract things irresistibly familiar, that makes a mathematical idea look like it was always already there. It was also about the similitude that a nimble protagonist of forward-looking international architectural culture like Friedman could tactically deploy to establish his work in settings shifting away from architectural form and its corollary image-making practices. Graphs afforded this felicitous opportunism: they could be images, made in Friedman's signature drawing style, and, when needed, they could pose as pristine mathematical objects safe from the perils of subjective biases and unchecked conventions.

For all the pleas to "see in a hard intellectual light," graph vision was not pure intellectual vision. It was still about producing and consuming images. Yet most skeptics of mathematical approaches to architecture steadily lambasted them for their ostracizing of aesthetic considerations. Amongst the most telling and admittedly colorful examples was the multipage repartee between conservatist aesthetic philosopher Roger Scruton and Philip Steadman, Dean Hawkes, and William J. Mitchell. This was triggered by Scruton's review of the 1976 LUBFS Centre work compilation *The Architecture of Form*, edited by March, which Scruton inflamingly titled "The Architecture of Stalinism."[166]

"After the war," Scruton wrote, "as London's Chief Architect and Planner, Martin was able to put his constructivist ideology into practice; and with the founding of the Cambridge Centre he ensured that the torch of Architectural Stalinism will be kept alight for as long as that institution continues."[167] The Centre, he continued, encouraged "a form of architecture which could fairly be described as totalitarian," guided by "a tyrannical censorship . . . over every aspect of building that does not have its numerical computation."[168] With the mathematics being unable to capture the complexity of all aspects of the building and without aesthetic considerations, which Scruton characterized as "inimical" to the constructivist's aims, one "acted blindly."[169]

Steadman, Mitchell, and Hawkes replied at length to the "miasma of misrepresentations and slurs."[170] In a section on "The Role of Mathematics," the authors argued that the visual education of architects, which Scruton seemed to prioritize, has always been mediated by mathematics such as descriptive geometry and perspective projection. "Simply the prediction of what a new building 'would be like' in the restricted sense of what its superficial visual appearance will be," they continued, "employs mathematical techniques—however 'transparent' and familiar these might be."[171] The new mathematics allowed a new kind of *envisioning*, not of what a building would look like visually but of *how it would work*.[172] Aside from Scruton's arguments' dubious political allegiances that seemed to negate architecture's social role, his critique also seemed to suffer elsewhere. "It does not take much acquaintance of even a passing kind with the basics of today's primary school mathematics," the authors scalded, "to recognise a Venn diagram; but apparently Dr Scruton's acquaintance has not got that far."[173]

The question for architects was not *if* mathematics, but *which* mathematics. As *The Geometry of Environment* reveals, the LUBFS Centre work was animated by postwar ruminations about the status of modern architecture as much as it was participating in the larger cultural and political project that was the new math. The "appreciation" for structure that we read new math educators aspiring to was manifest in the LUBFS Centre work and writings.

The new geometry would give architects a new apparatus for thinking about shapes at a time when architects' preoccupation with appearances seemed perilous and seemed to compromise disciplinary and professional legitimacy. At the interstices of the mathematical and architectural modernisms of which they were participants, LUBFS Centre researchers were not ready to do away with images. They endeavored to *draw out* architecture's hidden structures; those structures that school mathematics textbooks had so convincingly shown to be lurking in one's surround. Graphs accommodated, and in retrospect exposed, architecture's ambivalences toward seeing and drawing.

These ambivalences lie, chronologically and conceptually, somewhere between the immediacy of the drawing surface and the delegation of image-making to computations behind a screen. Unlike strokes, scribbles, etchings, and other depositions of material on a surface, digital images are electric flickers controlled by the symbolic calculus of computer scripts and programs or, more recently, the sorcery-like "denoising" of inscrutable neural networks. As architects grapple with the uneasy dependencies of architectural images on an abstract and hidden calculus, the story of graph vision focuses our attention away from the binary of surface and structure, idolatry and iconoclasm, and toward the cultural valencies of the technical regimes that reconcile them.

3 Tools

Ends

As we shift from graphs as *images* to graphs as *tools*, as instruments for picturing data and algorithmically automating design, we also move from representation toward abstraction, from epistemic toward operative claims, and from structure toward process. I use the preposition "toward" to point to orientations rather than stabilities, holding dear cautions against mistaking knowledge objects—here, for instance, mathematical diagrams, algorithms, and computer programs—for mere instruments.[2] Like their visualities, graphs' instrumentalities conjoin with epistemic quests and professional predicaments, with anxieties about the legitimacy of architectural decisions and the status of architecture as a modern discipline.[3] As tools, graphs served to support architecture and computing's foremost chimeric dream: the algorithmic automation of design.

The prospect of an "automated architect" has been stubbornly animating engagements of architecture and computing for decades.[4] Time and again, computer algorithms have been marshaled to automate not only rote architectural tasks and clerical work but also those cherished undertakings of configuring space and articulating form. Algorithmically automating architectural design has historically presupposed redefining it, to some extent at least, as a systematic *process* amenable to mathematical description. Could this be done for architecture? What were the motivations for, and implications of, putting treasured ideals of inspired intuition and creative genius under the rigors of mathematical analysis? What would recasting design as a set of explicit steps mean for architectural knowledge? And what would become of expertise and professionalism if "non-architects"—people or machines—could design through algorithmic recipes?

A discursive nebula of sorts, these questions, tainting as much as they are tantalizing, float above recurrent efforts to technically achieve the automatic production of architectural works. Today, machine learning coupled with a cornucopia of algorithmic techniques placed under the permissive umbrella of artificial intelligence is rekindling enthusiasm around the prospect of automatic design.[5] And while the armature of computational techniques marshaled toward this elusive goal expands and extends, graphs lurk as historic heralds of design as an algorithmic enterprise.

The fears, thrills, or often mundane inevitabilities of computer-driven automation were the backdrop of algorithmic impulses in architecture. Automation, in turn, as a technical and political project motivated processes of algorithmic abstraction, of hand-picking tasks deemed relevant for architectural design. This picking—and discarding—came together with, and as we will see were often conditioned by, processes of extracting data, information about contexts of architectural intervention.[6] Despite their professed autonomy and incontrovertible coherence, the algorithms for automatically generating architectural form were robustly rooted in specific architectural settings: that is, both in material practices of recording and abstracting human habitation of and labor in buildings, as well as in data-induced anxieties of architectural expert cultures that bartered concepts and techniques for reconstituting architectural work as an algorithmic process.[7]

Graphs of various kinds conceptually supported and technically enabled the illusory idea that architectural form could be automatically "generated" by processing empirical data based on stepwise mathematically articulable rules—on algorithms. As tools for holding together concrete data and algorithmic abstractions, graphs' instrumentalities were many, but they invariably involved processes of reconciling, even effacing, tensions between messy architectural settings and pristine design algorithms. Therein lies graphs' critical instrumentality: to retrieve tensions and undo these erasures; to put algorithms back *in architecture*—in concrete architectural settings rendered as data, in local architectural research contexts, and in disputes about architectural agency in an informationally infused world.

Data

In 1958, Sir Leslie Martin chaired a conference on architectural education organized by the Royal Institute of British Architects that would come to be recognized as a turning point in British architectural education. Known as the Oxford Conference, the event focused on entry and training standards for architects in relation to their desirable professional performance. At stake in the conference was the architect's responsibilities toward society or, as Martin put it, "the development of Architecture as a Public Service and what the public expects of the architect."[8] What may sound like yet another rumination on the architect's professional identity was in fact a pragmatic concern. It had to do with how architects should engage a rapidly expanding landscape of information on various aspects of buildings and cities.

With the introduction of social welfare policies after the end of the World War II, such as the National Health Service Act, the Education Act, and the New Towns Act, the UK had seen a rapid growth in public building with domestic, healthcare, educational, or governmental uses. Along with this focus, and a concern for using public funds efficiently, came the growth of various research agencies that focused on the collection of information on buildings and their inhabitants. These agencies included publicly founded and funded authorities such as the Building Research Station (BRS), formed in 1921 under the initiative of Minister of Health Neville Chamberlain with a mandate to advance efficient housing and construction methods, as well as charitable organizations such as the Nuffield Foundation. These organizations brought together experts from multiple disciplines in large-scale research projects that scrutinized building and construction in the UK, from the scale of major public works to that of the building detail. Events such as National Productivity Year in November 1962 under the initiative of the government-funded British Productivity Council boosted experimental projects for increasing efficiency in building design.

In the interdisciplinary settings of these momentous research projects, Martin remarked, architects came "at work" with many other parties: "structural engineers, mechanical engineers, production engineers, management and time study experts . . . clients, sociologists, psychologists, physicists and physiologists."[9] Architects' involvement in such large interdisciplinary teams exacerbated a growing sense of irrelevance. There was an increasing anxiety that architects were merely decorators of a technological environment that was being shaped in their absence. This was especially disconcerting for young students of architecture training to join the profession in the late 1950s and early 1960s. For instance, the first editorial of *The Architects' Journal* section dedicated to the British Architectural Students Association (BASA)—a consortium of student representatives from 25 British architecture schools founded in 1959—voiced confusion about their teacher's "big talk about art and technology," which "somehow sounds hollow when the one is so often used to embellish the other."[10]

Interlude 4

Consider these two scale drawings.

They are both floor plans of an operating theater suite, drawn circa 1963 in the UK. Both plans have a distinct relationship to data: the first contains it and the second is made by it.

The skeletal figure of the top plan looks like a graph, but it is something somewhat different: it is a "string diagram"—a device for recording and representing how workers move between tasks that they perform in specific locations of a building.[1] To make a string diagram, an observer would have to carry a scale drawing of the building's floor plan attached to a corkboard and follow a worker, in this example a nurse, as she went about her work. Every time the nurse stopped to perform a task, the observer would place a pin on the corkboard. Then the observer would use thread to connect the pins. This pins-and-thread model would then be used to make a drawing, by replacing the pins by points and the threads by lines. The more "trips" a nurse took between two locations, the more thread between the two pins and the thicker the line connecting them.

Now look at the drawing at the bottom. This is a new floor plan made by a human designer following instructions for the placement and dimensions of rooms outputted by a KDF9 English Electric Computer at the University of Liverpool running an Algol 60 algorithm for automatically generating floor plans. The floor plan might not look like much, but it represents a critical turning point. Before, computers had been used mainly to assign the locations of tasks in an existing floor plan or to compare the efficiency of different floor plan proposals for a given set of human movement data. This algorithm generated new floor plans. It maximized the efficiency of work operations by designing the architecture.

This algorithm was made possible by a subtle but decisive shift: stripping the string diagram of its metric aspects—the dimensions that tied it to the specificity of an existing floor plan—and transforming it into a list of points in space corresponding to tasks ("activities") and a list of relationships connecting these points ("trips").

The string diagram had to be turned into a graph.

a

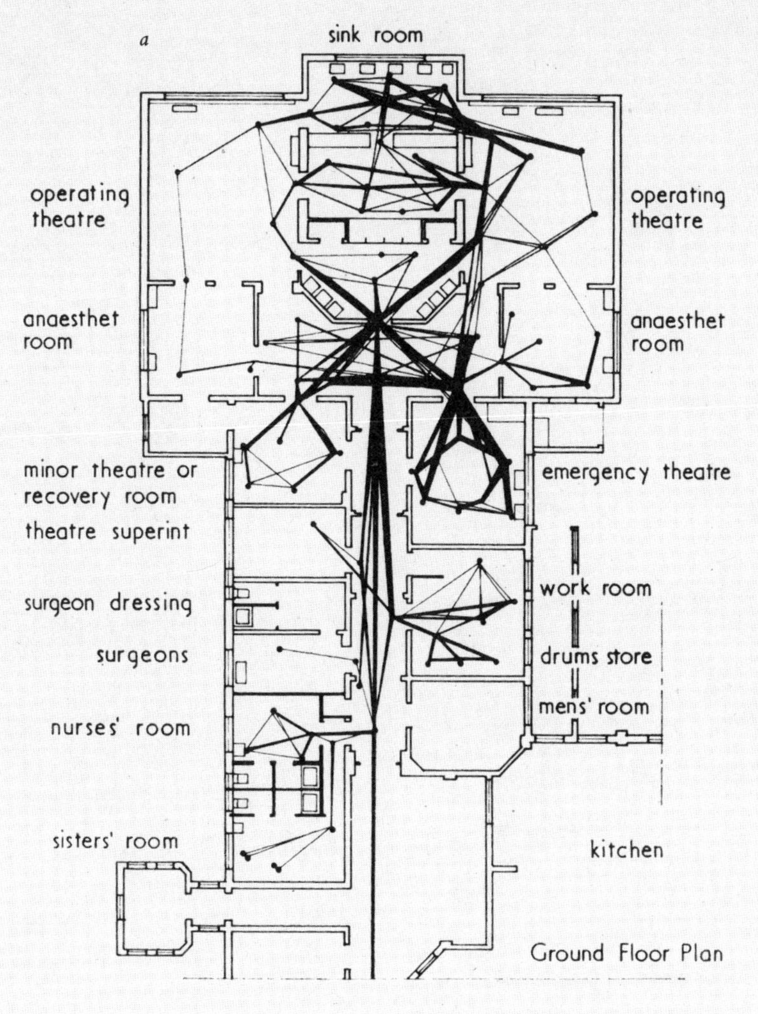

b

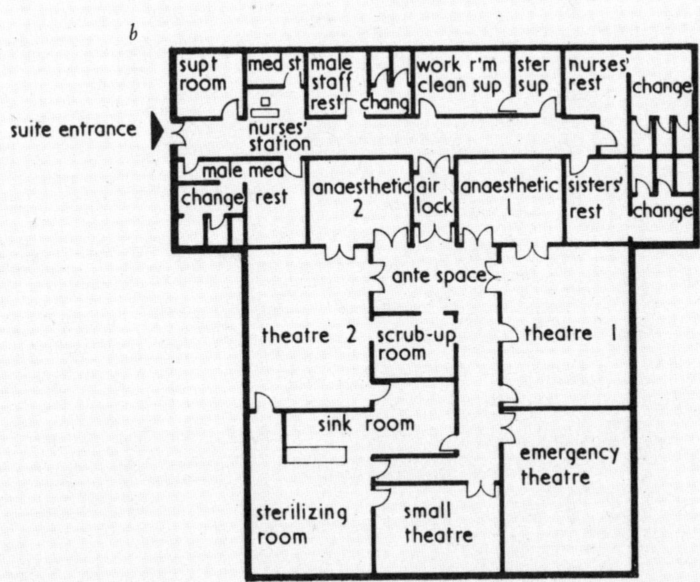

Figure 3.1
Source: B. Whitehead and M. Z. Eldars, "An Approach to the Optimum Layout of Single-Storey Buildings," *Architects' Journal* 139, no. 25 (1964): 1374 (*a*), 1380 (*b*). Courtesy of *The Architects' Journal*.

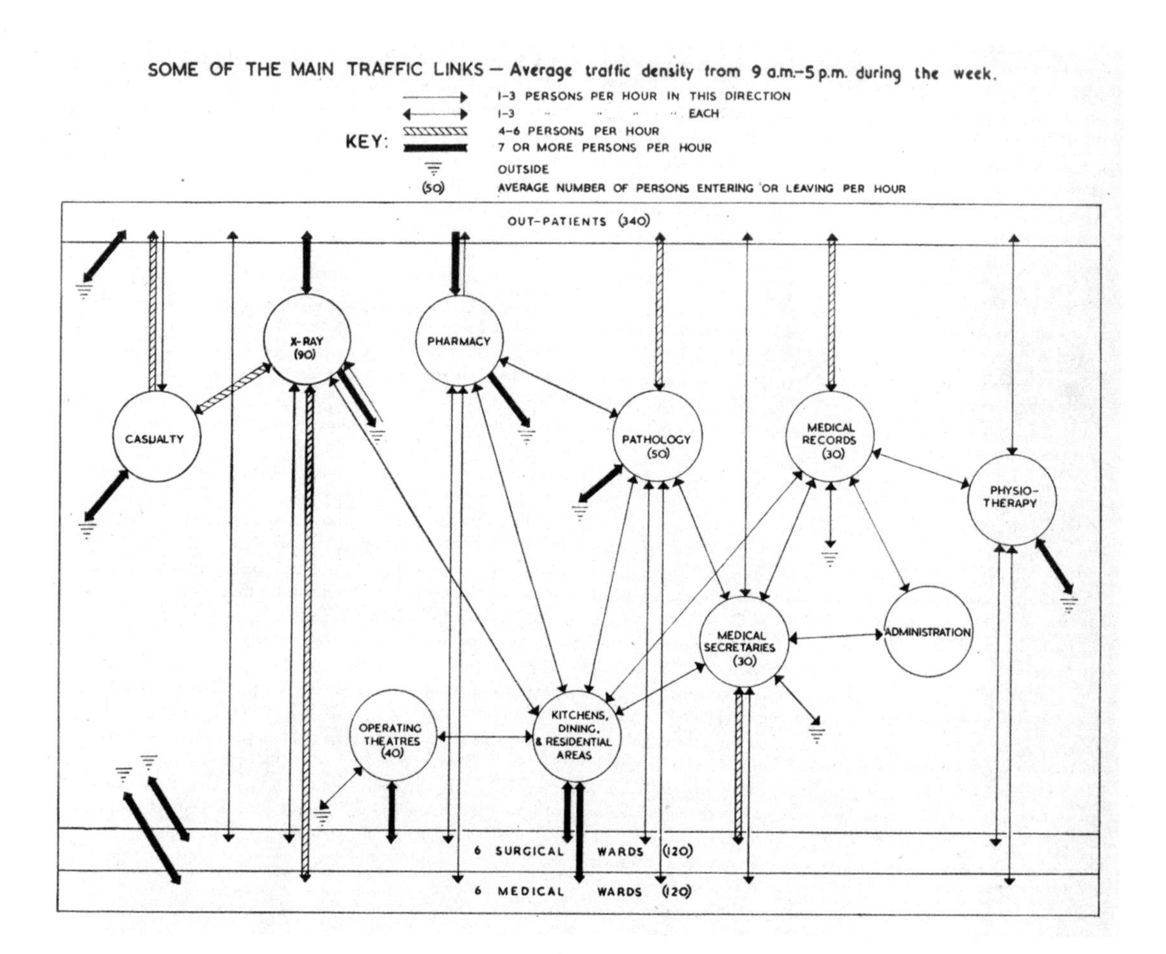

Figure 3.2
Diagram showing average human circulation among various hospital departments from 9 am to 5 pm during a week. The diagram is based on a study carried out by the Building Research Station in 1961–1962 in two large acute general hospitals of 600–700 beds, which was then expanded to nine acute general hospitals in England and Wales and recorded data for more than 54,000 trips. *Source*: M. Van Ments, "Hospital Planning: Internal Traffic in the General Hospital," *Architects' Journal Information Library* (1963): 29. Courtesy of *The Architects' Journal*.

Martin's student Christopher Alexander conveyed a similar sentiment in a talk that he gave along with Balkrishna V. Doshi in the 1962 International Design Conference in Aspen. Alexander and Doshi bemoaned architects' utter powerlessness:

> The chemist and physicist can manufacture entirely the materials . . . economic policy . . . is drafted by economists and politicians . . . the expert [in indoor climate] is the heating engineer . . . the experts in [the social repercussions of physical planning] are anthropologists and sociologists . . . [for] the psychological conditions . . . we will more likely consult the psychologists . . . [for] the economic effects of placing a building in one place rather than another . . . the economic planner is the expert. . . . And even in the simplest environmental problem of all, that of building dwellings, the builder rather than the architect reigns supreme. . . . So we are led to the strange conclusion that the architect at present plays almost no useful part in the creation of the environment.[11]

It was through "research"—a central keyword in postwar Anglo-American architecture—and not through professional practice that architects sought to reclaim their authority in shaping the environment.[12] To do so, it seemed urgent to find ways to use massive amounts of observational data that large-scale studies of various building types yielded, not only to *evaluate* existing buildings and cities—evaluations often revealing blatant inefficiencies—but for the *generation* of new and better ones. Mathematics, carrying connotations of rigor, reason, and exactitude, offered techniques for this purpose by carving out a new area of architectural operation: the devising of *processes* for integrating and operationalizing information collected by outside experts in the context of architectural design.

At once a demand and an opportunity, the prospect of moving from the analysis of a set of design requirements to the description of a proposed physical thing—be it an object, a floor layout, an urban master plan—ignited optimism about a new definition of design that reliably incorporated and complied with available information; a design that would be, as some might say today, *data-driven*. Rather than being created as a work of artistic genius, form would be attained through rule-based, algorithmic processes parsing available data collected in ongoing research pursuits. The obscurity and messiness of decisions that a designer was seen as traditionally making when designing would be replaced by a transparent set of steps defined through mathematics and logic, ultimately advancing a new disciplinary identity for architects and designers. Alexander's polemic in his early article "The Revolution Finished Twenty Years Ago" is telling of this attitude:

> Certainly, it's always fun to play soldiers. But there's no time for fun when there is serious work to do. The work of this generation, and of succeeding generations will be the work of refinement. Work that must proceed slowly, doggedly, methodically. There are no longer battles to be fought.[13]

Some 12 years later, Berkeley architect and urbanist Roger Montgomery would recognize Alexander as the author of the "first manifesto" of a "worldwide movement" to "modernize design methods and bring scientific rigor into [designers'] ancient craft."[14] Alexander was a key figure for joining mathematics and computer applications in architecture with a *method* for ordering information about design and translating them into the organization of physical form. His design method, saliently based on graphs and published in the influential 1964 book *Notes on the Synthesis of Form*, was taken up by several architects and planners internationally, who adapted it for educational experiments or developed computer implementations.[15] Despite its unique impact and influence, amplified by Alexander's polemical writings and prolific publications, his method cannot be separated from a broader turn in the design disciplines, writ large, toward goal-directed, procedural definitions of design.

Such definitions were shaped under the umbrella of "rational design methods," an influential movement that spanned across different disciplines concerned with "design"—mechanical, chemical, and electrical engineering, industrial design, urban planning, and architecture. Design methods sought to overhaul the infamous caricature of design as a mix of precedent, convention, rules of thumb, and subjective judgments. They displaced "creativity" from a nebulous ideal having to do with inspired intuition to a technorational definition tied to achieving far-reaching goals through a structured process. Design methods recast design as a goal-seeking *process* amenable to mathematical analysis and articulable in an explicit series of steps. This produced a new disciplinary focus for design, from the physical form of the final artifact (object, building, city) to the steps and decisions that led to it.

Design's and architecture's postwar affiliation with science and rationality continued a lineage of early twentieth-century movements and institutions such as De Stijl and the Bauhaus, while also aligning with lateral developments in engineering education.[16] For example, a landmark conference on engineering education held in September 1960 at the Case Institute of Technology in Cleveland, Ohio, advanced definitions of design as an iterative decision-making process of optimally converting resources into desired ends.[17] This revamped definition would place design at the center of engineering education and enable the integration of advanced analytical methods in the development of engineering systems and devices. Publications such as Morris Asimow's 1962 *Introduction to Design* disseminated analytical methods such as morphological analysis, also advanced by Swiss astronomer Fritz Zwicky, for breaking down the design into smaller parts or phases and evaluating alternative "solutions" within each phase.[18] It did not take long for these adjacent debates and techniques to make their way into architecture and industrial design with path-setting events such as the Conference on Systematic and Intuitive Methods in Engineering, Industrial Design, Architecture and Communications organized in 1962 at Imperial College London, and the founding, in 1966, of academic organizations such as the Design Research Society in the UK and the Design Methods Group in North America for the promotion and dissemination of research on design processes.

Looking back at the history of the systematic study of design in architecture, design methodologist Geoffrey Broadbent argued that despite conceptual kinship between design methods and disciplines such as cybernetics or operations research, the field's foundations should be sought in new mathematical techniques that became available to designers.[19] "The new maths, with a certain amount of statistics," he wrote, "has been almost as influential in the development of new design methods as all the other sources and disciplines put together."[20] In other words, Broadbent seemed to suggest, it was specific mathematical techniques that made concepts such as goals, feedback, and means-ends analysis applicable in design. Similar statements echoed repeatedly in design-methods-related writing, from a reviewer of the 1962 Conference on Design Methods musing about the promises of "topological nets" presented by conference co-organizer J. C. Jones and participant Christopher Alexander, to academic engineer William Spillers opening a major 1974 symposium on design theory by casting graphs as fundamental tools for forging a "common ground" for all of design by treating it as a "generalized network problem."[21]

The short-lived but avid engagement with rational design methods left a legacy of new discourse, academic programs, conferences, printed media such as journals and newsletters, but also of new *images* of the design process. Diagrams showing the flow of information from one decision to another, spirals looping between "analysis," "synthesis," and "evaluation," trees breaking down the design process into manageable parts pictured new topologies of design decisions. Despite their different appearances and divergent structures, representations of the design process being advanced within design methods asserted design as an informational process that followed a decision-making program. This program could be represented as something akin to a computer program flowchart.[22]

These representations also laid the groundwork for the convergence of systematic design with computer tools for design's automation. In debates about the systematization of design processes, computers and algorithms came to symbolize what design researcher Nigel Cross characterized as "the truest form of systematic methods almost by definition."[23] Describing a design process in machine form mandated its unforgivingly systematic structuring. Early algorithms for design automation emerged at the nexus of concerns with systematic design processes and the growing availability of information, of data, about building. Both anathema and opportunity, the sheer amount of information triggered sentiments of "bewilderment," as Alexander famously wrote in the *Notes*, but also compelled the data's manipulation.[24]

Layouts

In the early 1960s, building scientists and architectural researchers appealed for the use of massive amounts of observational data that large-scale studies of various building types yielded not only for the evaluation of existing buildings and cities—evaluations that were often scathing for their inefficiencies—but for the *generation* of new and better ones. In this wishful translation, digital electronic computers that sparsely appeared at that time in universities such as the Atlas Computer in Manchester and in computer research centers such as the Ferranti Research Center in London bore hope and possibility.

Without any interactive or graphical capabilities and occupying entire rooms, computers were still perceived as elaborate calculators. They were seen as able not only to take away the tedium of rote and repetitive tasks but also, as a 1962 *Architects' Journal Information Library* technical article titled "Computers in Building" stated, to "handle and process calculations of bewildering complexity at a speed far in excess of that of the human brain."[25] At the time that article was written, the main applications of computers to building were PERT (program evaluation research techniques) and CPM (the critical path method). These methods using digital electronic computers to tackle the scheduling of construction operations and manage complex building projects that required labor and materials.

The critical path method, for example, broke down the building process into distinct operations and represented the process in the form of a graph. The graph's points represented the operations and their associated attributes such as time and cost and the lines represented relationships of dependency between the operations (which operations needed to be completed for others to proceed). The computer would calculate the minimum duration for the complete project and the operations that were critical for this to be achieved: the "critical path" through the graph. This helped make decisions on which operations to prioritize with more workers or money and examine various alternative scenarios based on modifications of each operation's time and cost. Managerial and logistical in focus, these methods established the computer as "the executive's tool *par excellence*."[26] Computer-designed buildings, the author remarked, were "still very much in the realms of science fiction," as it was not clear how to formulate the work and decisions of architects on a mathematical basis that would make them amenable to computer analysis.[27]

By May 1963, however, the *Architects' Journal* was announcing an experiment in which a computer had been put to work to plan a hospital, rendering what had been fiction a few months before as a real yet financially precarious possibility. Seeing "computers in action" in architecture seemed to have, as the journal announcement suggested, three possible results: "complete failure, complete success or some indeterminate mixture of the two."[28]

The "Computers in Action" piece preluded a longer article in the "Computers in Building" series that detailed an experimental use of a computer technique for planning a hospital. The demonstration was

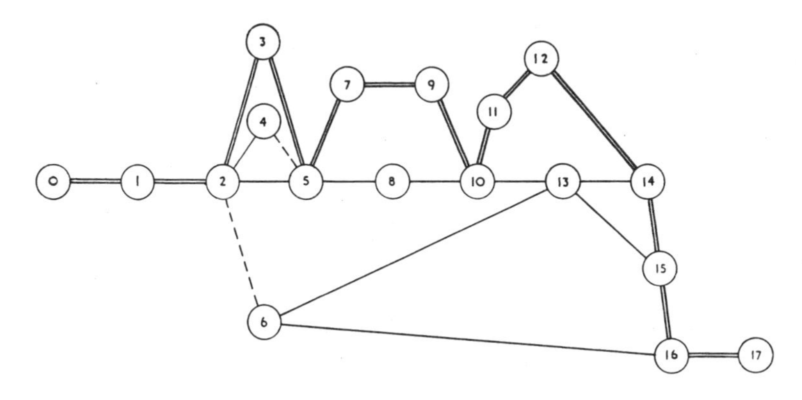

Figure 3.3
Critical path diagram illustrating activities (lines) and events (points) indicating the sequence, direction, and duration of each activity. *Source*: "Computers in Building: PERT and CMP," *Architects' Journal Information Library* 136, no. 24 (1962): 1332. Courtesy of *The Architects' Journal*.

under the auspices of W. E. Tatton Brown, chief architect at the Ministry of Health, and was directed at architects and other ministry officers.[29] Bruce Archer, one of the ten organizers of the 1962 Conference on Design Methods and subsequently a major figure in the institutionalization and dissemination of design methods, spearheaded the experimental demonstration. The study was funded by Michael Farr (Design Integration Ltd.), where Archer was employed at the time following two years as a visiting scholar at the Hochschule für Gestaltung in Ulm, where he had been exposed to operations research and related mathematical methods.[30] In London, Archer led a research project on the design of hospital equipment, such as standard hospital beds, funded by the Nuffield Foundation. The demonstration presented in "Computers in Building" did not focus on equipment, but instead on planning the hospital based on room schedules that indicated rooms' various purposes and servicing needs.

Archer collaborated with Ferranti Ltd., one of the United Kingdom's leading electrical equipment companies, which had also established a computing center in London. Conducted as part of Ferranti's service to architects, the experimental demonstration proposed digital electronic computers as aids in grappling with the informational maze that was the modern hospital.[31] The experiment used data from a large study by the Building Research Station that took place between 1961 and 1962 in 11 large acute hospitals and recorded data for tens of thousands of "trips"—movements between distinct locations in the hospital where various activities occurred.[32] With the National Health System investing a large amount of public funds in the redesign and modernization of hospitals, planning new hospitals in congruence with the information that was amassed appeared imperative.[33]

In planning the hospital, architects were confronted with copious amounts of information in the form of room schedules, based on which they articulated the hospital plan. Each configuration came with different efficiencies, and it was impossible to tell whether the architect had picked the best plan among the myriad possibilities, of which they only evaluated a handful. For only 16 activities, the author of the report of the Archer and Ferranti demonstration counted 1,309,111,372,800 possibilities that required trial and evaluation: "the dupe," the article stated, casually invoking a common troublesome metaphor, "at which that fast-working slave, the computer, is best used."[34]

But using computers required two fundamental steps: translating data into numbers and finding the right mathematical model. The approach that Archer proposed was the adaptation of a computer programming technique known as the "transportation technique." The paper detailed a remarkable translation—a *displacement*—from calculations on the economic use of tankers, oil wells, and refineries to the plotting of efficient routes for vehicles transporting goods between various points to the hospital. The key innovation, the author stated, was to turn the transportation technique "inside out" and use it to optimize the locations of the points of origin instead of the movements between them.[35]

Archer reconceptualized the hospital as static activities and communications between them, corresponding to the movement of people, services, material, and information. Each activity pair was given a priority factor based on the importance of the communication between them and the need for their proximity. Then Archer proposed a "topological model" based on the expected shape of the building, which he rendered as a three-dimensional square grid whose vertices corresponded to the number of activities to be allocated. The computer placed activities on the vertices and then calculated the total "cost" of the configuration by multiplying the priority of each activity pair by their "distance" (the number of edges between the two activities) in the abstract grid. The optimal solution was the one with the lowest total cost.

The technical study closed with the announcement of the next step: a program written for a hypothetical hospital of 1,000 rooms and a quest for an architect willing to offer his project as "the world's first guinea pig for trials of the technique."[36] "At first sight the enormity of the task is frightening," the article stated, "but of course, once done, the principles used to decide priority and distance values will form a ready guide to all later problems."[37]

The presentation of the Archer and Ferranti demonstration project in *The Architects' Journal Information Library* stirred reaction from B. Whitehead, a lecturer in the Department of Building Science at the University of Liverpool. In a letter to the journal, Whitehead criticized Archer's method for being indifferent toward the intricate interrelations among all the different activities that make up the hospital.

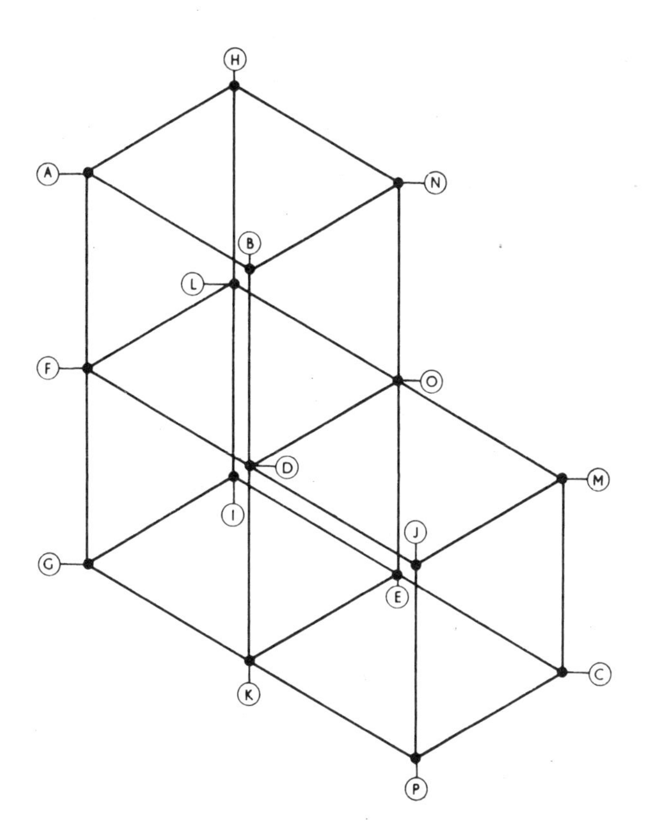

Figure 3.4
Archer's topological model of the hospital with assignments of activities to locations (nodes). *Source*: "Computers in Building: Planning Accommodation for Hospitals and the Transportation Problem Technique," *Architects' Journal Information Library* (1963): 142. Courtesy of *The Architects' Journal*.

This risked absurd and untrustworthy outputs, such as optimizing the outpatient department and the X-ray department in relation to the operating theater only and placing them at opposite ends of the hospital despite their close relationship.[38] Whitehead announced a new approach to this issue and a computer program that he and his students were in the process of developing and testing. "We believe," he wrote, "that the programme will be reasonably economical in computer time and will give, within the limitations of the accuracy of the original data, a close approximation to an ideal layout."[39] Immediately after the letter followed a note from the editors: "We would be very pleased to hear more about Mr Whitehead's alternative approach."[40]

A few months later, writing with PhD student Mohamed Zakaria Ahmed Eldars, Whitehead published a computer program written in Algol that could compute the best location for 37 activities in ten minutes on a KDF9 English Electric Computer.[41] The demonstration case presented in the article involved an operating theater suite consisting of 21 rooms. The algorithm was based on data representing hospital staff's movement in the operating theater in the form of string diagrams: physical thread-and-nail models that represented movement between activities taking place in distinct locations on the hospital floor plan. It is likely that the data was drawn from the work of the

a

total journeys		
117	1	sisters' changing room
171	2	nurses' changing room
717	3	surgeons' rest room
399	4	surgeons' changing room
46	5	superintendent's room
24	6	medical store
395	7	small theatre
376	8	anaesthetic room no 1
711	9	theatre no 1
528	10	sink room
488	11	sterilising room
677	12	scrub up room
1115	13	ante-space and nurses' station
711	14	theatre no 2
376	15	anaesthetic room no 2
395	16	emergency theatre
254	17	workroom and clean supply
146	18	sterile supply room
249	19	male staff changing room
546	20	nurses' station
305	21	the entrance

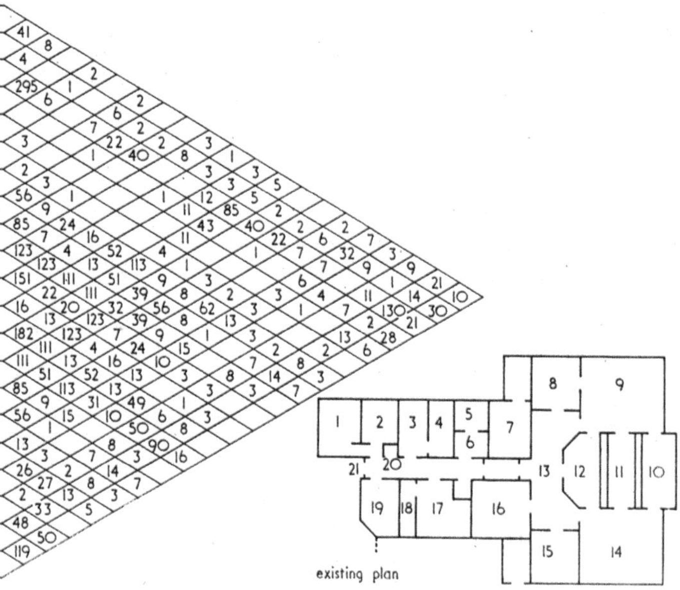

existing plan

b

number of element		ante-space, nurses station 1	2	3	4	scrub up 5	6	theatre no 1 7	8	9	10	11	12	theatre no 2 13	14	15	16	17	18	sink room 19	20	21	22
ante-space, nurses' station	1																						
	2	1000																					
	3	1000	1000																				
	4	1000	1000	1000																			
scrub up	5	23	23	23	23																		
	6	23	23	23	23	1000																	
theatre no 1	7	5	5	5	5	10	10																
	8	5	5	5	5	10	10	1000															
	9	5	5	5	5	10	10	1000	1000														
	10	5	5	5	5	10	10	1000	1000	1000													
	11	5	5	5	5	10	10	1000	1000	1000	1000												
	12	5	5	5	5	10	10	1000	1000	1000	1000	1000											
theatre no 2	13	5	5	5	5	10	10	1	1	1	1	1	1										
	14	5	5	5	5	10	10	1	1	1	1	1	1	1000									
	15	5	5	5	5	10	10	1	1	1	1	1	1	1000	1000								
	16	5	5	5	5	10	10	1	1	1	1	1	1	1000	1000	1000							
	17	5	5	5	5	10	10	1	1	1	1	1	1	1000	1000	1000	1000						
	18	5	5	5	5	10	10	1	1	1	1	1	1	1000	1000	1000	1000	1000					
sink room	19	2	2	2	2	4	4	6	6	6	6	6	6	6	6	6	6	6	6				
	20	2	2	2	2	4	4	6	6	6	6	6	6	6	6	6	6	6	6	1000			
	21	2	2	2	2	4	4	6	6	6	6	6	6	6	6	6	6	6	6	1000	1000		
sterilising room	22	1	1	1	1	3	3	6	6	6	6	6	6	6	6	6	6	6	6	16	16	16	
	23	1	1	1	1	3	3	6	6	6	6	6	6	6	6	6	6	6	6	16	16	16	100[partial]
	24	1	1	1	1	3	3	6	6	6	6	6	6	6	6	6	6	6	6	16	16	16	100[partial]
small theatre	25	1	1	1	1	9	9	1	1	1	1	1	1	1	1	1	1	1	1	3	3	3	2[partial]
	26	1	1	1	1	9	9	1	1	1	1	1	1	1	1	1	1	1	1	3	3	3	2

area 600 sq ft i e 6 elements

i e this figure represents number of standard journeys between elements 6 and 8

dummy relationship

Nuffield Provincial Hospitals Trust Job Analysis team that had performed such studies since 1949.[42] However, the specific hospital floor plan and string diagram used by Whitehead and Eldars were not featured in the Nuffield Trust influential guide *Studies in the Functions and Design of Hospitals*.[43]

The data that the algorithm processed had to be prepared manually based on the string diagram information for each individual type of staff, for example a doctor or a nurse. One first identified pairs of activities in the operating theater and then calculated how many trips each type of staff member took between these endpoints. One then needed to multiply the result by the total number of staff members of this type employed in the hospital and then weigh it based on each member's salary scale to calculate the "cost" of each trip. The total number of trips was used to make a matrix called an "association chart." The header row and header column of the table listed all the activities that needed to be placed in the operating theater suite. Each table cell carried information about a trip between the activity listed in its row and the activity listed in its column. The cell contained the total weighted trips between these activities, representing the connection strength between each activity pair. To account for variations in the square footage needed for activities, each activity was assigned a number of 10 ft by 10 ft "elements" or units of floor area. Using this, Whitehead and Eldars produced what they called the "relationships matrix"—a square grid of cells, with each cell corresponding to one "element." Each grid cell had a unique store number in the computer memory.

This is where the computation began. The algorithm assigned activities to grid cells with the aim of keeping close together the activities connected with the largest number of trips. The computer program's output was a printout of a list of "element" numbers with their positions in the grid. The architect's task would be to translate the algorithmic output into architecture: to assign room functions to cells of a gridded page using the printout as a guide and then turn that schematic diagram into a floor plan by adding wall thicknesses, doors, and windows.

Among the references that Whitehead and Eldars cited were various methods for the systematic planning of factories and production facilities developed by management scientists and industrial engineers. These references included Richard Muther's "systematic layout planning" method, a set of representational conventions and procedures for identifying, visualizing, and rating the various activities, relationships, and alternatives involved in a project of laying out equipment in an existing building.[44] Whitehead and Eldars not only invoked factories rhetorically as a model for a well-planned, efficient hospital but also used many of the conventions included in these methods in

Figure 3.5 (*opposite*) "Association chart" (*a*) and "relationships matrix" (*b*) accounting for the area needed for each activity. *Source*: B. Whitehead and M. Z. Eldars, "An Approach to the Optimum Layout of Single-Storey Buildings," *Architects' Journal* 139, no. 25 (1964): 1375 (*a*), 1376 (*b*). Courtesy of *The Architects' Journal*.

Figure 3.6 (*opposite*)
Part of the printed computer output indicating the locations of activities per area element, translated manually to a "locations matrix" and transformed into a diagrammatic layout. *Source*: B. Whitehead and M. Z. Eldars, "An Approach to the Optimum Layout of Single-Storey Buildings," *Architects' Journal* 139, no. 25 (1964): 1379. Courtesy of *The Architects' Journal*.

their algorithm. Leafing through the *Systematic Layout Planning* manual, for example, one finds illustrations of the factory planner, a white male engineer, drawing an "activity relationship diagram" to "picture" "flow-of-materials" and "other-than-flow" relationships by overlaying a transparent grid sheet on the factory floor plan. The planner is then depicted moving around physical square pieces of cellulose or colored plastic on a grid fitted on a drawing of the factory floor plan to assign area to these activities.

In Whitehead and Eldars's work, the material square pieces moved around with one's indices map onto the elemental units of the relationships matrix, assigned new positions not with fingers but with electric signals. Their computer program automated some of the material manual practices involved in methods of systematic layout planning.[45] However, the program also added a layer of mathematical optimization to this automation. Unlike concurrent efforts to use digital electronic computers to plan hospital layouts in the United States, some of which focused mainly on evaluation of different layout alternatives proposed by human designers, the Whitehead and Eldars program algorithmically automated the generation of optimal layouts for a specific set of data.[46] The hardware configuration of the KDF9 computer that did not support interactive displays, and a distinct research ecology that focused on computers' mathematical and algorithmic as opposed to interactive capabilities, forged distinct research orientations in the United Kingdom such as these reflected in Whitehead and Eldars's work.[47]

The Whitehead and Eldars algorithm became a standard reference in early research on computer-aided architectural design.[48] This was not because of its results but because of a hidden abstraction that lay at its core. While Whitehead and Eldars drew and spoke of string diagrams and matrices, they did not speak of graphs. Yet it was precisely the treatment of the string diagram as a graph—a representation of pure relationships without metric measurements—that made it possible to even consider a data-to-form translation. The string diagram kept track of healthcare workers' movement data and tied the information to the scale and dimensions of an existing floor plan. This link had to be cut for a new floor plan to become possible. Whitehead and Eldars freed human movement data from the fixed geometry in which these had been recorded by transforming string diagrams into a set of points and dimensionless lines. Now recast as graphs, string diagrams could be translated into matrices in a one-to-one way. Matrices could easily be entered as input into a computer that then algorithmically reshuffled the activity-location pairs they represented to output more efficient floor plans.

Among the many subsequent publications that referenced the Whitehead and Eldars article was a series of working papers published

```
PROGRAMME NO. 24-B
ELEMENT LOCATION.

+1.70000000,    +001      Location No. 17
+4.00000000,    +001      Element No. 40

+1.80000000,    +001
+5.50000000,    +001

+2.30000000,    +001
+2.40000000,    +001

+2.40000000,    +001      (each pair of
+1.60000000,    +001      numbers con-
                          sists of an
+2.50000000,    +001      element number
+1.50000000,    +001      with its chosen
                          location)
+2.60000000,    +001
+1.80000000,    +001

+2.70000000,    +001
+3.90000000,    +001

+2.80000000,    +001      Example of
+4.20000000,    +001      computer
                          output
+2.90000000,    +001
+5.40000000,    +001

+3.20000000,    +001
+2.30000000,    +001
```

8 *Computer output*

9 *Completed locations matrix*

		54	44	43	46	47	48		
	55	42	41	38	35	45	49	50	
	40	39	37	36	33	34	51	52	
		18	17	2	3	11	12	53	
		15	13	1	4	7	9		
		16	14	6	5	8	10	32	
		24	27	19	20	25	28	30	
			23	21	27	26	29	31	

10 *Diagrammatic layout derived from the computer*

superinten-dent room 54
male staff changing and rest room 44 43
work room and clean supply 46 47
48
medical store 55
entrance 42
nurses' station 41
38
35
sterile supply rm 45
nurses' changing and rest room 49 50
medical staff change 40
medical staff rest 39
anaesthetic room no 2 37 36
anaesthetic room no 1 33 34
sisters' changing and rest room 51 52
general theatre-1 18 17 15 13 16 14
ante-space 2 3 1 4
general theatre-2 11 12 7 9 8 10
53
scrub-up 6 5
32
sterilising room 24 22 23
sink room 19 20 21
25
emergency theatre 28 30 29 31
small theatre 27 26

Figure 3.7
Configuring a layout by moving small square pieces that correspond to area units. *Source*: Richard Muther, *Systematic Layout Planning* (Boston: Cahners Books, 1961), 8–11. Courtesy of Richard Muther & Associates.

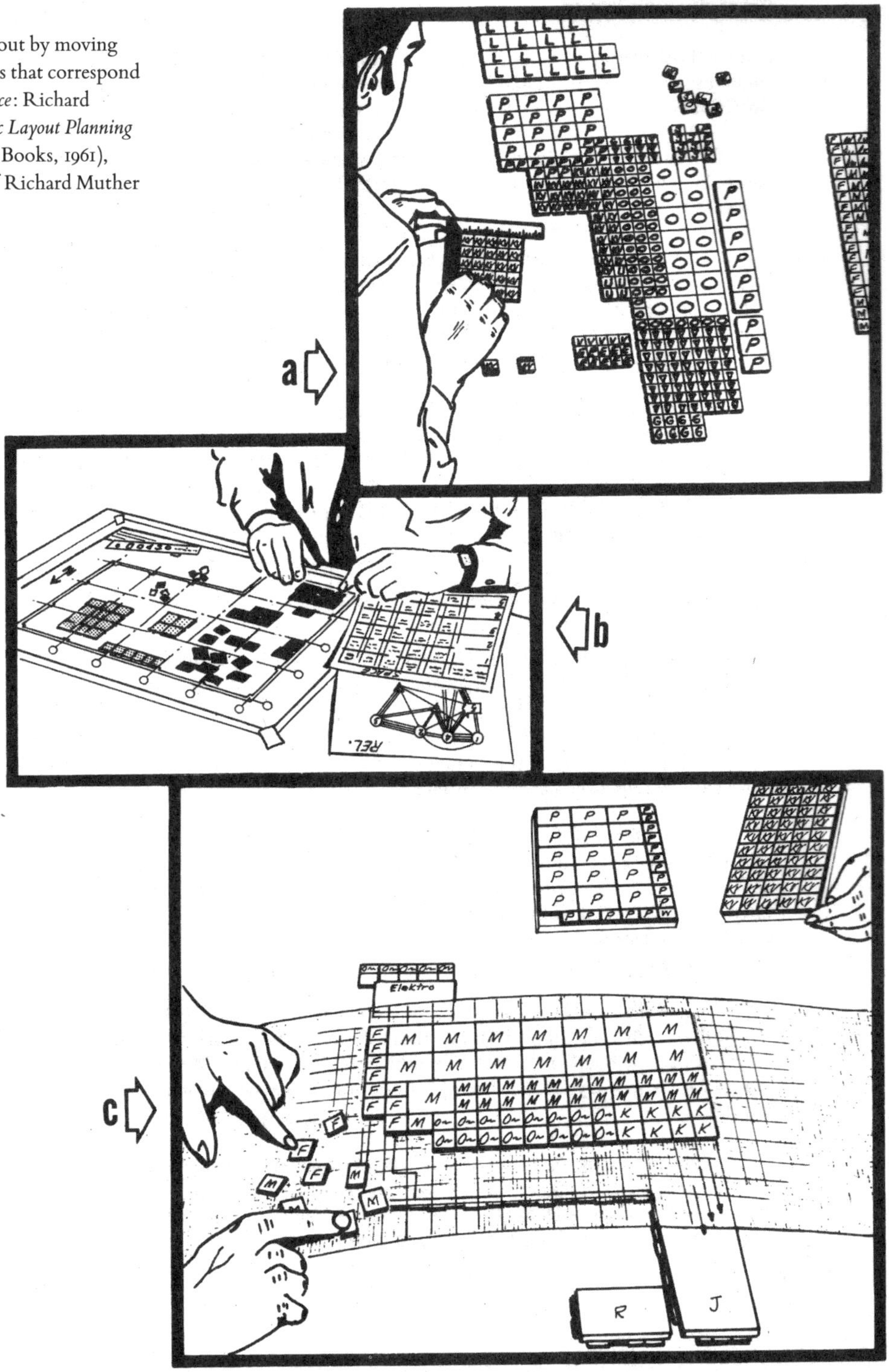

between 1969 and 1970 by the LUBFS Centre to summarize and propose computer-aided design methods for office buildings. The working papers were written by Philip Tabor, leader of the LUBFS Centre's Offices Study group that had started its operation in 1967 with funding from the Ministry of Public Building and Works. Around the same time, the ministry had instituted a standing Committee on the Application of Computers in the Construction Industry (CACCI), with the goal to review and recommend steps toward the advancement and consolidation of computer applications in the construction industry.[49] Despite not contributing significant applications for architectural design, at least as a working paper by the Architecture and Building Aids Computer Unit (ABACUS) at the University of Strathclyde in Scotland was reporting in 1970, CACCI helped consolidate and promote research on computer-aided architectural design through conferences and bibliographic compendia.[50]

After 1969, the Ministry of Public Building and Works sponsored a Computer Aided Design Study group within the LUBFS Centre in support of the Offices Study. The Offices Study group divided its contract from the ministry into two substudies: one concerned with "physical evaluation" (lighting, sound, air circulation, et cetera) of the office building, led by Dean Hawkes, and one concerned with the building's "design and spatial evaluation," led by Tabor.[51] The design of large office buildings involved many different parties and was notoriously inflexible, with bad decisions being expensive to unmake.[52] Tabor set out to "lubricate" the process by finding ways to correlate relevant data and processes for making design decisions.[53] In order to achieve this, it became necessary to forecast quickly and accurately the effects that various decisions would bear on the functioning of an office. The pathway for doing this was mathematical models of office buildings. Mathematical models were, as we have seen, LUBFS Centre's modus operandi. The 1971 *Architectural Design* editorial-manifesto of the Centre put it clearly:

> Our common method is to formulate mathematical models which enable us to explore, experimentally, ranges of spatial patterns which accommodate various activities. . . . On the one hand, the work requires us to find appropriate mathematical representations which are isomorphic to the spatial and physical form of the building, site, or urban area; and on the other the modelling of patterns of activities at these scales.[54]

Construing space and activity as distinct but interrelated *systems*, LUBFS Centre researchers set out to model structures of space and structures of activity. Graphs were an ideal tool for doing that. In several LUBFS Centre models, graphs were used to describe both spatial structure and activity patterns and to establish correspondences between these otherwise incommensurable domains. The end goal of such mappings was a particular kind of efficiency: minimizing the costs not by which an office building was produced, but instead by which it *functioned*. The architects' failure to account for the building's "operational efficiency" or "costs in use" led to financial loss and hampered productivity.

With arguments similar to Whitehead and Eldars's about the advantages of human movement information as being amenable to empirical observation and quantification, Tabor suggested using circulation patterns as the main data for the design of an office building layout. In four consecutive working papers, he reviewed techniques for modeling pedestrian circulation, locating activities in space, analyzing communication patterns in an organization, and evaluating routes in a building.[55] Some of these techniques were also featured in *The Geometry of Environment*, with Tabor offering advice to Steadman on material to include in chapters 12 to 14.

On page 11 of LUBFS Centre Working Paper 17, Tabor presented the string diagram shown in the Whitehead and Eldars paper redrawn as a graph. The two drawings looked almost identical. The main difference, aside from a 90-degree rotation, was the supersized circles that indicated the points—the activities—that formed the departures and arrivals for nurses' and surgeons' trips mapped onto the hospital floor. The oversized points refocused visual attention from the string diagram's lines, representing circulation, to the graph's points: a structure of activity-location pairs that could be lifted from the specific floor plan and recombined. This differentiated the graph not only from the operating theater suite string diagram, but also from visual representations of human movement that architects had been familiar with since the interwar period—from Alexander Klein's late 1920s "graphical analysis" methods for the evaluation of apartment floor plans to circulation diagrams accompanying building floor plans by second Bauhaus director Hannes Meyer and his students, to the activity diagrams that building scientists were encountering in their filiations with industrial engineers. These graphical methods were all analytical. The representation of human movement was attached to a set geometry. The points became handles by which one could pull the geometry around. Graphs were a transformable and transformative abstraction: a set of points that could be reorganized in multiple, but countable, ways, with each reorganization implying a different floor plan geometry.

At the same time, the visual similarity of the graph to established visual and material practices in architecture such as the string diagram, the graphical method, or modernist circulation diagrams validated its

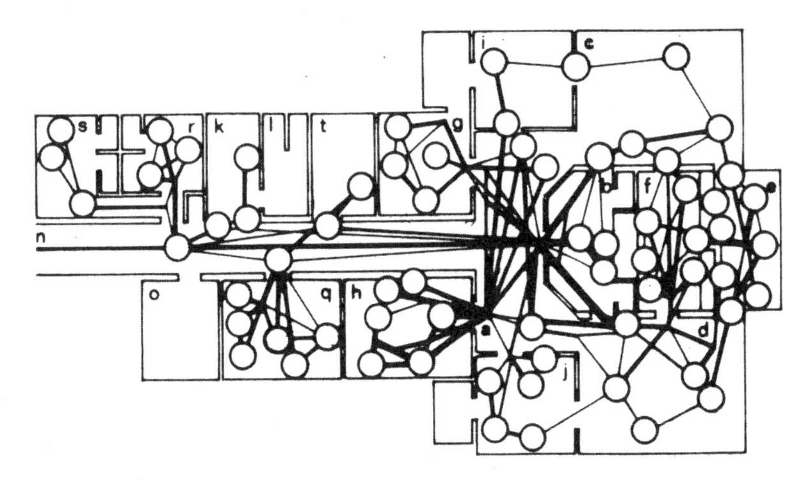

Figure 3.8
Graph showing a nurse's daily movements around a hospital operating theater suite, adapted by Philip Tabor from Whitehead and Eldars's 1964 article "An Approach to the Optimum Layout of Single-Storey Buildings." *Source*: Philip Tabor, "Pedestrian Circulation in Offices," *LUBFS Working Paper* 17 (1969), 11. Courtesy of the Martin Centre for Architectural and Urban Studies.

use. Employing the graph did not entail importing an alien mathematical language into architecture. Rather, it meant imbuing familiar representational practices with a surplus of mathematical potency. The graph being visually like the string diagram mattered as much as it being conceptually and mathematically distinct. This drawing negotiated the relationship between the graph as a visually manipulable tool and the skeleton as a hidden calculative device.

Tabor's working paper reports were rife with statements about the aesthetic and epistemic values of graphs. For instance, in contrasting three types of graphs commonly used to diagram office organization—"organizational" graphs (representing relationships of subordination and responsibility), "classificatory" graphs (representing relationships of containment), and "operational" graphs (representing connections between origin and destination points in the office)—Tabor remarked that operational graphs were of superior usefulness. "The connections are described more comprehensively and in more detail than in any of the others," he explained; "they are *real and not conceptual* [emphasis mine], and they are those whose efficiency is directly affected by the building."[56] By appraising the "realistic" aspects of the operational graph, Tabor was pointing to the one-to-one mapping of the graph's points with empirically apprehensible entities that in turn corresponded to distinct locations in space.

Among techniques reviewed and adapted by Tabor, graphs figured prominently as a tool for describing, and mapping between, the office's operational and spatial structure, and, crucially, for generating this spatial structure. Apart from participating in the office operational organization, activities also had a "spatial identity," a unique location representable by a dimensionless point.[57] Because of this one-to-one matching between activity and location, ordering activities through

Figure 3.9
Isomorphisms between the spatial structure of an office, its pedestrian traffic, and its managerial structure. Total number of daily trips between coworkers (*a*), "association graph" showing total number of daily trips between each coworker pair (*b*), and structure-of-command graph as a partial graph of the association graph (*c*). *Source*: Lionel March and Philip Steadman, *The Geometry of Environment: An Introduction to Spatial Organization in Design* (Cambridge, MA: MIT Press, 1974), 291 (*a*), 293 (*b*), 298 (*c*). Copyright Philip Steadman and the Estate of Lionel March.

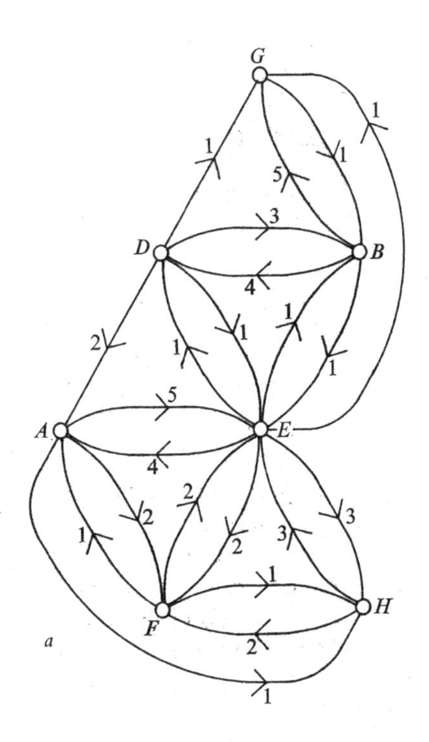

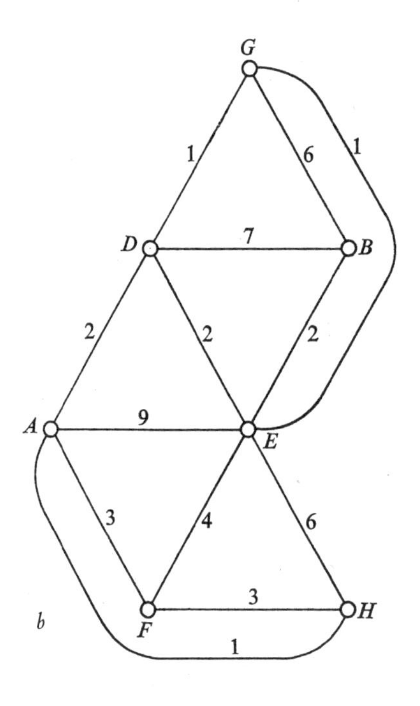

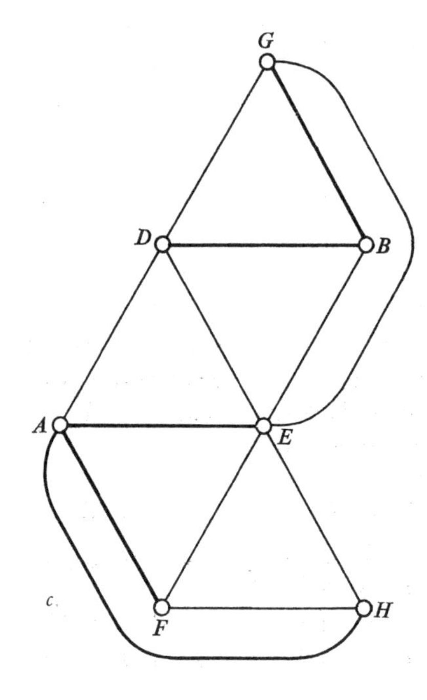

pedestrian circulation bore the potential of also ordering the physical spaces they occupied.[58] As we saw with Whitehead and Eldars, once activities had been assigned to locations, it was possible to evaluate the traffic between them by multiplying their "distance" and their "association," as well as to use similar measures to generate optimal layouts for a specific set of data. These algorithms, Tabor suggested, could be performed either manually, with pencil and paper, or automatically, with digital electronic computers. Automatic techniques could handle three times as many activities, but usually required predefining the shape of the building and, importantly, lacked "the flexibility of manual procedures."[59] Pencil and paper methods, all graph theoretic in origin, on the other hand, presented a "simplicity rightly discouraging rigidity in their use."[60]

This statement echoed a dispute about showing and hiding that had ensued in the *Architects' Journal* letters to the author section. This was about four years before Tabor's report on graphs as generative tools for architecture. In October 1964, a month before the dispute, Peter Levin, a researcher working in collaboration with the Building Research Board, had published an article claiming that optimum layouts in buildings such as hospitals could be achieved "without resorting to a computer and with the aid only of a paper and pencil."[61]

Levin used the data from Whitehead and Eldars's paper, arguing that it sufficed to draw this data as graphs and use simple rules from graph theory to derive an optimal layout. He listed numerous advantages of the method, including the designer having "complete control over the process" and the method not requiring any equipment other than the tools architects were already using.[62]

Whitehead and Eldars, their paper being the sole reference in Levin's article, replied. Graphs' simplicity, they wrote, would likely attract the attention of architects. But this simplicity contained the seeds of danger: the complacent would assume they were already doing what the graphs can do and the more "open-minded" would assume that graphs were already doing what computers can do. Following a list of refutations, Whitehead and Eldars concluded: "It is still possible to wash clothes in a dolly tub, but it is faster and cheaper in the long run to use a washing machine."[63]

Values of immediacy and malleability attached to graphs appeared to circumvent bewildering complexities of data or complicated computations performed at the obscure interiors of electronic computers. Graphs could be trusted to establish mappings between a pattern of activities and a pattern of space, but they were also easy to understand and manipulate, making them a prime tool for visually trained architects. And while Whitehead and Eldars sermonized on computers' calculative supremacy and their symbolizing of technological progress

Figure 3.10
Translation of graph into spatial layout (left) and graph representations of different room enclosures. *Source*: P. H. Levin, "The Use of Graphs to Decide the Optimum Layout of Buildings," *Architect's Journal* 140 (1964): 814. Courtesy of *The Architects' Journal*.

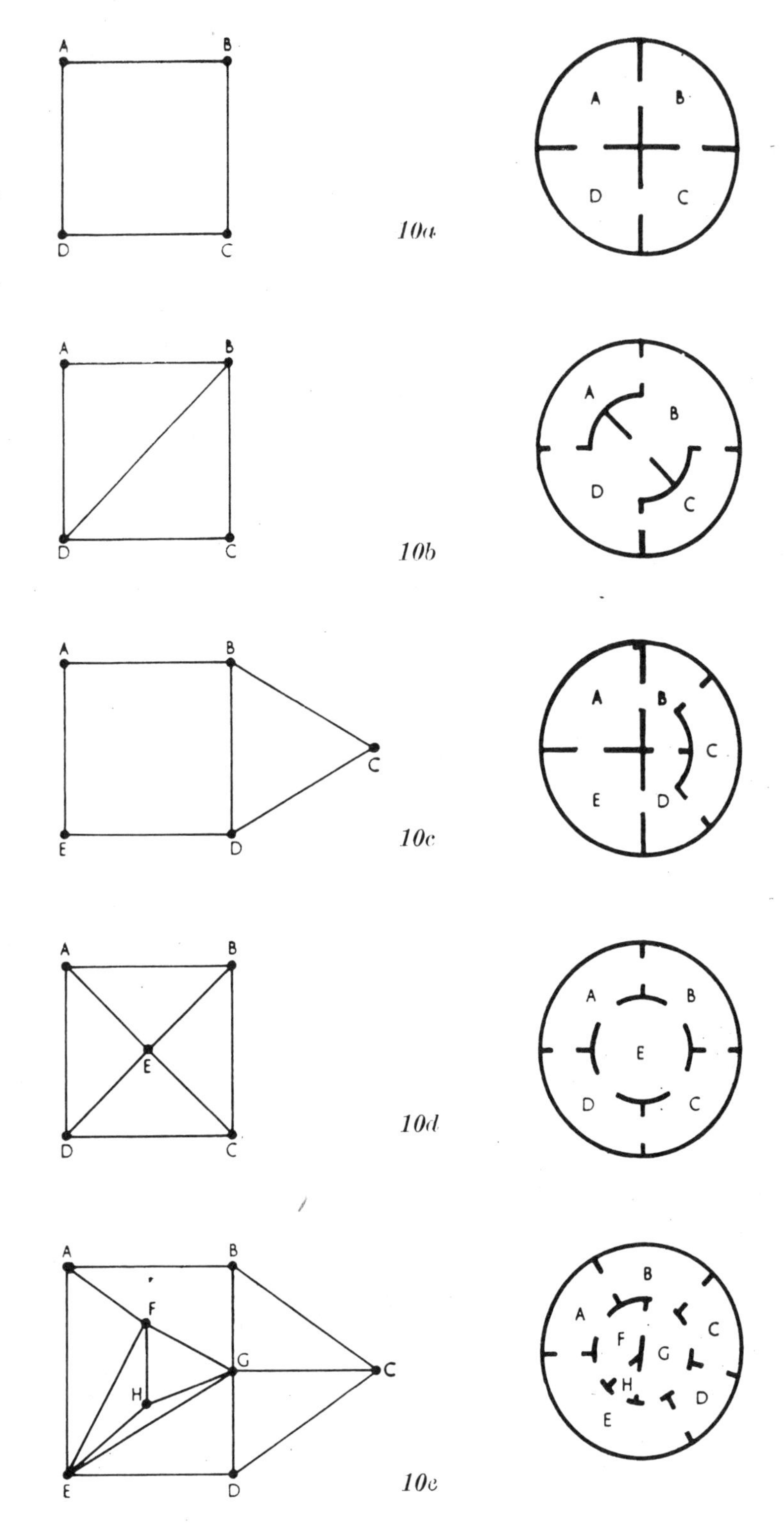

and modernization, many architectural researchers found an irresistible appeal in the visual, paper quality of graph-based processes. Either as the conceptual foundation of a computer algorithm in Whitehead and Eldars's computer program or drawn on paper in Levin's method, graphs recall other data inscriptions—the string diagrams, the activity patterns, and their material renderings with thread or pencil lines on vellum.

As mathematical tools, graphs—or their hidden skeleton counterparts—were instrumental in configuring a particular definition of generative design: shuffling relationships between the graph's points to produce alternative building layouts and evaluating them based on data about the activities that these layouts ought to accommodate. The prospect of undistorted translations from patterns of activity to patterns of space was an exhilarating one, which dominated the first 15 years of research in architecture, mathematics, and computers under the label of "space allocation" or "automatic spatial synthesis." However, not everyone was on board.

In a famous condemnation of computer-aided architectural design, Christopher Alexander argued against the usefulness of examining large numbers of layouts. In his presentation at the 1964 Architecture and the Computer conference organized at the Boston Architectural Center—one of the first conferences dedicated specifically to computers and architecture—Alexander chastised growing efforts to "apply" computers to architectural design for being "misguided, dangerous, and foolish."[64] Also published in the journal *Landscape* with the title "A Much Asked Question about Computers in Design," Alexander's diatribe rejected automatic layout programs.[65] By generating layouts based on criteria that could be "measured and encoded" (for example, circulation costs), Alexander bemoaned, these programs led to the ironic situation "that the very tool which has been invented to unravel complexities imposes such severe restrictions on the design problems it can solve that the real source of complexity has to be eliminated before the tool can even get to it."[66] Alexander argued that the examination of many alternatives was reasonable in theory. However, although the computer could examine many alternatives, these alternatives were constrained to a restricted type of solution.[67] Because of its inability to produce "truly unexpected" alternatives, the computer outputted design options that were different only with respect to trivial measurements and failed to be qualitatively different.[68] Alexander likened this process to comparing millimeter differences in the dimension of a block of wood intended to block a car from sliding or "measuring the size of a cooking apple with a micrometer."[69]

Interlude 5

Could the data itself dictate the conceptual categories by which architecture should be designed?

Alexander asked that question as an employee of the Harvard-MIT Joint Center for Urban Studies working with Serge Chermayeff, who also advised his doctoral dissertation, on a research project called "The Urban House." The project sought "a vocabulary capable of describing the infinite variety of elements, situations, activities, or events that make up the complex organism 'house'."[70] Alexander was aware of similar attempts to derive new conceptual categories under which to classify "factual data," such as Knud Lönberg-Holm and Theodore Larson's Development Index, but dismissed them as "a little awry."[71]

The data itself, he argued, contained a nonarbitrary order.

He worked out this idea by proposing a new category for organizing urban house data: the "failure."[72]

"Failure" denoted a kind of physical event that prevented a need from being satisfied (for example, sleep prevented by bioclimatic discomfort). "Failures" established relationships between design requirements and aspects of physical form. They also established "interactions" between the requirements themselves. Sometimes failures shared data, other times they were corrected by the same operations, and other times the correction of one failure aggravated the other. Similar relations of overlap, reinforcement, or conflict were then established among the failures' corresponding requirements.

By considering "the relations themselves, or links" between failures, it would be possible to compute "a working programme for design": both a structure, representing a logical organization of design requirements, and a process, indicating in what order the designer should tackle them.[73]

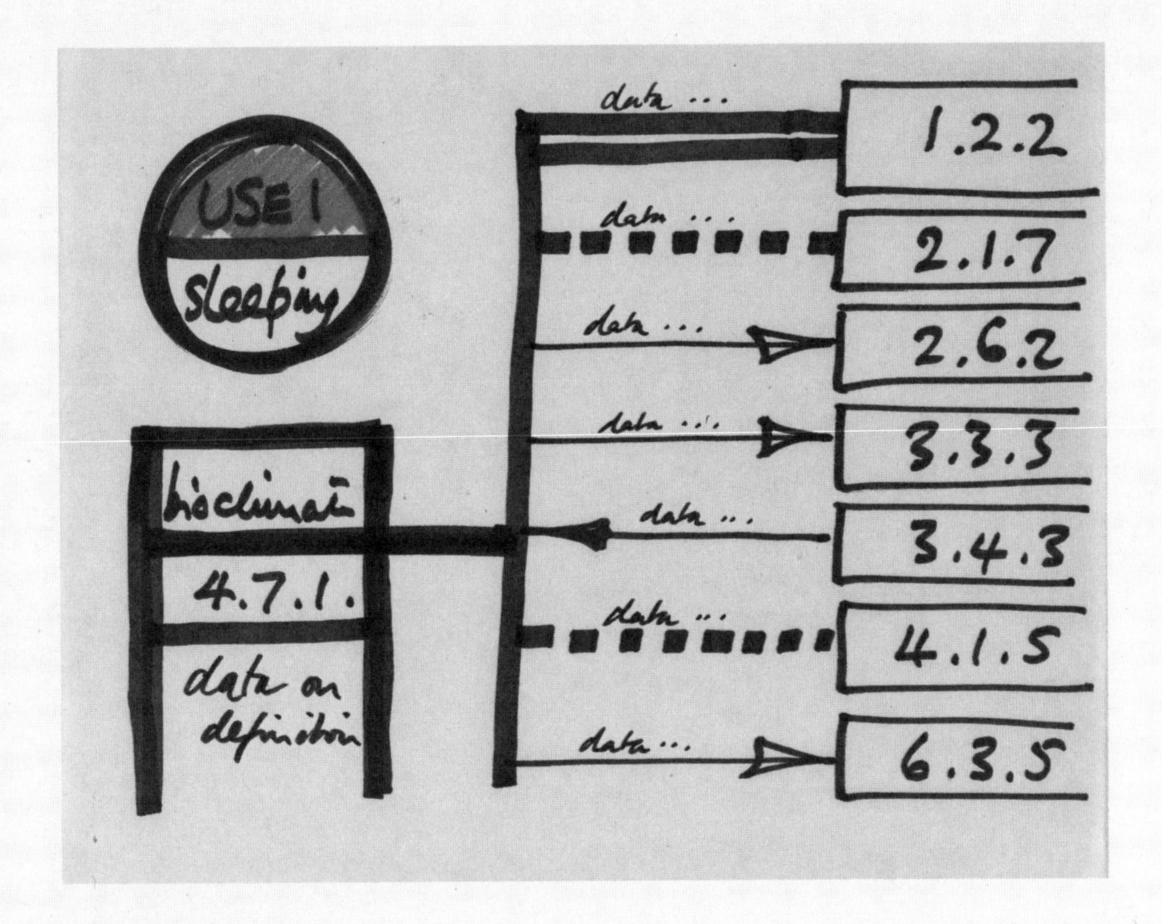

Figure 3.11
Source: Christopher Alexander, "Letter to Chermayeff Re: Failure Cards," 1960. Box 4, Folder "Alexander, Christopher, 1958–1966." Serge Chermayeff architectural records and papers, 1909–1980, Avery Architectural & Fine Arts Library, Columbia University. Courtesy of the Avery Architectural & Fine Arts Library, Columbia University.

Patterns

By the time that Alexander presented his denunciation of the wrong-headed uses of computers in design, he was already reaping recognition as a leader in the use of mathematical and algorithmic techniques in architecture. Indeed, he was among the first architects to develop a computer application intended to aid designers with the ordering of a brief's design requirements and the sequencing of design decisions for meeting that brief. HIDECS 2, as the program was called, a short for HIerarchical DEComposition System 2, was developed during Alexander's doctoral studies at Harvard while he was pursuing a consultancy at the MIT Civil Engineering Systems Laboratory, in collaboration with Laboratory member Marvin L. Manheim.[74] Running on the IBM 709 of the MIT Computation Center, under the control of the Fortran Monitor System in use at the Center during the second half of 1961, the system was a computer implementation of a graph-based theory of design that Alexander had been developing for his doctoral dissertation since 1958.[75]

Alexander's theory, published as *Notes on the Synthesis of Form* in 1964, grouped design requirements into a hierarchical *tree* (a graph where two points are connected by exactly one line) after analyzing relations of incompatibility among these requirements, which Alexander called "misfits."[76] The tree essentially broke down a complex set of interconnected requirements into smaller, more manageable subgroups while ensuring that these subgroups were as independent as possible from one another. The human designer then was responsible for producing what Alexander called "diagrams," partial sketches of the design that addressed these subgroups. In the final step of the process, the tree was used to compose the diagrams back together to produce the final design.

The method was based on a preoccupation with ordering information that was becoming available through research on mass-produced housing. Alexander became exposed to these questions during formative collaborations with Chermayeff at the Harvard-MIT Joint Center for Urban Studies. A key concern for Chermayeff and Alexander was to come up with conceptual "components" for categorizing the elements of a house that came from *within* the information itself. Rather than talking about houses in terms of kitchens, living rooms, and bathrooms, it was relationships emerging from the analysis of empirically collected data that would give rise to these conceptual components.

Alexander was a firm supporter of the idea that any set of data already carried an internal, nonarbitrary, logical order that could be brought to the surface through mathematical analysis.[77] This structure could then organize the requirements of a particular design "problem" to which this information corresponded. Requirements would be represented as graphs and then, using a mathematical method for decomposing complex graphs into simpler ones, ordered into hierarchical trees.[78] These trees represented the structure of information relevant for a design problem while also showing designers the best path for addressing these requirements—this would be in the form of an ordered stepwise process that Alexander called a "program."[79] This was the foundation of the *Notes*.

Apart from pioneering mathematically powered design methods in architecture, Alexander is also famous for rejecting his own mathematical methods.[80] Alexander's move away from the Notes' "tree" happened precisely when his early calls for rationality were becoming slogans for researchers and scholarly organizations internationally concerned with systematic design methods. The next major work in his career, *A Pattern Language*—published in 1977 as the outcome of more than a decade of research at a Berkeley-based nonprofit corporation that Alexander founded with his students Sara Ishikawa and Murray Silverstein—is patently different from the *Notes* in rhetoric, style, readership, and goals. Because of its accessible format and underlying formal ideas, *A Pattern Language* gained popularity among amateur designers and homeowners, was emulated in software design, received skepticism from architectural theorists and critics, was applauded by the counterculture, and was celebrated as a systematic tool for participatory design. And yet between these seemingly opposing poles of the *Notes* and *A Pattern Language*, between the Harvard Graduate School of Design and the University of California, Berkeley, between technical rationality and counterculture, lies a story of using graphs to negotiate messy data and pristine algorithms.

In 1965 Alexander took a two-year leave of absence for visa-related reasons from the University of California, Berkeley, where he had been hired as an assistant professor in 1963, and temporarily moved to London. At the beginning of his leave, Chermayeff—newly appointed department head in Yale University's School of Architecture—invited Alexander to join the faculty there.[81] Alexander wrote back declining the position with the excuse of his many open projects and dependent students at Berkeley. He also updated Chermayeff on his ongoing concerns:

> I am giving a lot of thought, just now, to the question of integrating all the various systems that make up the whole urban system—and how they interlock. This is particularly important, since no one person can ever work on all of it—and since individuals, and individual projects, must concentrate on parts only, it becomes critical to be sure that all these parts will fit together.[82]

To make sense of this statement we need to consider an ambitious research program that Alexander had put in place when he came to Berkeley. At the time, the University of California, Berkeley was undergoing major institutional reforms, including the restructuring of architecture, landscape architecture, and planning departments into a newly established College of Environmental Design. Alongside this change came an increased demand for systematic research in architecture and landscape architecture that could match the already established research tradition in planning.[83] Alexander's mathematical background, research experience, and involvement with design at all scales of the built environment—qualities that Chermayeff had outlined in an enthusiastic recommendation letter to Berkeley's Architecture Department chairman Charles Moore—aligned with that department's unfolding agendas.[84]

Soon after his arrival, Alexander wrote a research proposal titled "Ten Year Program for Research on Environmental Design."[85] There he outlined his plans for a colossal undertaking: "the problem of designing the form of the entire urban environment."[86] "Despite its deeply theoretical hard core," Alexander projected, the program would appeal to a cornucopia of individual agencies and would potentially ultimately acquire the character of a service or consultancy organization.[87] The program's aspiration was "to create, over a period of ten years, a conceptual design for the modern city."[88] The aim was to devise new "components" (physical and conceptual) that would come to replace the conventional categories by which designers were thinking about buildings and the city. "The search, in each individual project," Alexander wrote, "is for a system of requirements which is so coherent and independent that its implications for the form of the environment can be studied in isolation."[89] Each of the subsystems would be published as a separate book with pertinent data collected in a "permanent information store."[90]

Alexander envisioned the program as a "federation of individual basic projects," with new "basic projects" emerging every so often from the analysis of empirical data using the graph-based methods that he had developed during his years at Harvard.[91] Although the research program's conceptual substrate and the way that the basic projects were derived and linked to each other had its basis in abstract mathematical reasoning, the basic projects themselves would be relentlessly empirical. Alexander wrote:

> All the projects will have one method in common: the minute observation of details, aimed at getting a clear picture of the demands, needs, requirements, and stresses which arise in a problem area. Although some formalized methods of observation do exist, it is intensity of observation, and attention to detail which will yield results, rather than exactly defined methodology.[92]

When he submitted the proposal, he had already completed a "basic project" to show as proof of concept: the conceptual design of Bay

Area Rapid Transit (BART) district stations in San Francisco. The work was performed as a consultancy to the firm of Wurster, Bernardi & Emmons, founded in 1945 by William Wurster, former School of Architecture dean and instrumental figure in the creation of Berkeley's College of Environmental Design. The BART project employed three principal investigators and two assistants. The project ran for eleven months with a total cost of $140,000.[93]

Alexander started working on the BART project almost immediately after his move to California, in collaboration with Ishikawa and Berkeley student Van Maren King. The team was soon after joined by Michael Baker and Patrick Hyslop. Over the course of eleven months the team performed extensive research into and visits to numerous transit systems to collect 390 requirements.[94] These were then "interacted" through HIDECS to produce the sequence by which Alexander and his team were to group the requirements, produce schematic diagrams for requirement subgroups, and combine these diagrams—a process similar to the *Notes* method.[95] The process proved arduous and with limited practical significance, as HIDECS ended up breaking the problem down into rather obvious components (subsystems), such as entrances, ticketing platforms, and so on.[96] However, the BART study was productive in other ways. The sheer labor put into the BART diagrams' invention and the gradual linking of human behaviors with physical form cultivated the idea of reusing some of the BART diagrams as generic entities for the design of rapid transit stations irrespective of geographic location or specific requirements.

A turning point in Alexander's "Ten Year Program" took place during his two-year leave in London. This is a less-known period in his career: most descriptions of Alexander's activity in the UK, where he undertook his mathematical and first architectural training under Leslie Martin starting in 1953, usually stop at his condemnation of the uncoordinated, nonsensical, and "absurd" status of architectural education and his rejection of a position at the Building Research Station to instead move to Massachusetts and pursue doctoral studies at Harvard.[97]

Alexander's visiting appointment at the Offices Development Group of the Ministry of Public Building and Works in the mid-1960s is a critical juncture for our story. This is not only because the appointment brought Alexander into the context of the traffic in buildings studies and space allocation work that we saw previously, but also because the Offices Development Group appointment played a crucial role in the development of *A Pattern Language*. It was during that appointment that Alexander solidified and stabilized his distance from other projects contending to generate form through observational data on people and cities—work that, as we saw in the previous section, the Ministry of Public Building and Works continued to fund well into the 1970s. During this appointment, Alexander elucidated a distinction between the oft-used concept of "activity" and his definition of a "pattern." This distinction carved a path for the use of graphs as generative devices that led all the way to *A Pattern Language*.

Alexander expanded on the idea that some of the diagrams produced in specific design studies such as the BART project could be generalized to similar building or infrastructure types. He worked on developing "environmental rules" for an "urban rule system," a system he outlined in a paper which he presented in Basel titled "The Coordination of the Urban Rule System."[98] Rules were essentially ways to combine diagrams that had been severed from the specificities of a unique design brief and generalized to apply to the design of similar "subsystems" of a city. Unlike what we saw with Whitehead and Eldars or with Tabor, graphs would not generate form by *mapping* activities and matching them with spatial structures. Instead, the graphs' points would represent human-form relations as units of a larger structure—what Alexander and his colleagues labeled "environmental structure." This structure would serve as the foundation of a *generative language*—a set of ordered rules—that would allow one to design without disturbing the order of the environmental structure.

The Ministry of Public Building and Works, as we have seen, was a seedbed in the UK for the development of design methods, architectural research, and, in the late 1960s, computer-aided design. Under the headship of Ian Moore, the Offices Development Group had done extensive work on collecting information about the human occupation of buildings and outputted a method for architectural programming called the "Activity Data Method."[99] In keeping with the systematic layout planning techniques that inspired computer programs for automatic floor plan generation, Moore and his collaborators used "activity"—actions taking place in discrete locations—as a unit of architectural programming, to replace the more passive concept of "need."[100]

During his appointment as a research architect in the Offices Development Group, Alexander collaborated with architect and social researcher Barry Poyner, who was working on user requirements at the Office, to develop the so-called "relational theory." The theory was published at Berkeley as a report titled *The Atoms of Environmental Structure*.[101] In the report, Alexander and Poyner stood critically against the Offices Development Group's concept of "activity." Their criticism stemmed from the observation that "activity" described actions *separate* from their physical environment. Remember the visual rhetoric of Tabor's operating theater drawing: activities, represented as the points of a graph, could be detached from the architectural plan that contained them. In fact, this independence from an existing layout was perceived as their main advantage: by reordering activities based on criteria of efficiency one could then generate new, more efficient floor plan configurations.

Alexander and Poyner saw activities as intricately linked with their physical settings. A new floor plan, in other words, would fundamentally change the nature of activities. To activity, they juxtaposed the concept of a "tendency."[102] Like activities, tendencies also stemmed from systematic observation of people in physical environments. However, Alexander and Poyner argued tendencies to be irreducibly *relational*. They proposed an intriguing formulation: a tendency described the needs that people tried to satisfy when a physical arrangement gave them the opportunity.[103] We might for instance say that people have the tendency to sit down when there are seats available in a public transit vehicle. Tendencies were statistical; they referred to behaviors that *most* people tended to adopt in a particular setting.

Coming out of methods initially proposed in industrial engineering and management science, the concept of activity carried with it some of the logics of the factory floor. There, tasks were repetitive and mechanical. They were part of a work sequence that was conceived independently of the environment in which it took place. Whitehead and Eldars conceptualized the work of medical staff in a hospital in similar ways. And because techniques come with their own inertias, Tabor adopted this understanding of work in proposing methods for analyzing and designing offices. The idea of a tendency, on the other hand, implied a degree of open-endedness in how a building's occupants acted. Although inarguably behavioristic—in the sense that it focused on observable behaviors and assumed those to be triggered by an external stimulus—the concept of a tendency enabled broadening Alexander's theory beyond specific institutions in which human actions were regimented by predictable work protocols and task assignments, out into the "wild" of an urban context.[104]

Having defined the concept of a tendency, Alexander adapted ideas from the *Notes* to suggest how tendencies should be treated. The *Notes* were about mathematically ordering the requirements of a design brief by analyzing conflicts between these requirements. The graph (be it tree or semilattice) ordered the requirements in such a way that such conflicts could be eliminated. In keeping with this thinking, Alexander and Poyner strove to eliminate conflicts between people's tendencies—where "tendencies" replaced the *Notes*' "requirements." In order to achieve this, they defined the concept of a "relation," which Alexander later called a "pattern." A relation was a geometrical arrangement that had the property of preventing such conflicts among tendencies.[105] *The Atoms of Environmental Structure* included an example of entrance relations for suburban houses that Alexander developed together with Ishikawa.[106]

In the *Atoms* report, Alexander and Poyner posited that "environmental structure" was composed of atomic "relations." Environmental structure would become intelligible by investigating how relations "interlocked."[107] The BART study, together with subsequent work on relations that Alexander pursued at the Offices Development Group, helped articulate a new keyword that would play a short-lived but important role in the development of Alexander's structural thinking: the "relational complex." The "relational complex" was essentially a concept for matching physical relations between elements with functional relations between requirements in a building. It was based on the idea that underpinning each building's form was a set of physical relations that "control the way that buildings work [underlined in the original]."[108]

The concept appeared in a 1966 article coauthored by the BART project team (Alexander, Van King, Ishikawa, Hyslop, and Baker) and titled "Relational Complexes in Architecture." In an introduction to the article, the journal editor described the work as seeking "to make use in architectural design of the *new mathematics of relationship* [emphasis mine] and the capabilities of the computer, while at the same time remaining fully cognizant of the complexities and subtleties that are an essential part of all architecture."[109]

Indeed, in their text the authors extensively promoted the "transition from number to structure" and reprimanded other factions engaged with systematic design methods for their bias toward numerical exactness. Such attachment to numbers "obscure[d] basic relationships" and cultivated a false sense of precision, leading to complacency.[110] The crux of the article was the establishment of a structural mapping between a physical relation (a positional or topological relation of physical elements) and a functional requirement. If a common physical element was found in two or more requirements, then the relations were characterized as "interlocking." The "relational complex" was a sophisticated structural entity describing how various simple physical relations interlocked. "Design," the authors aphorized, "is the invention of relational complexes. We must learn to define them, and to design them."[111]

"Relations" between physical form and human behavior formed the "atoms" of "environmental structure." Research on relations was not unlike Alexander's "basic projects" that he had envisioned upon his arrival at Berkeley in 1963. The definition of each "relation," or to use its subsequent name, "pattern," required an extensive project of data collection on some part of the urban environment. In this respect, defining patterns was also continuous with the empirical program of cataloging and ordering data about design requirements that Alexander undertook during his PhD work leading up to the *Notes*. Aside from being in keeping with a long-standing inquiry into the relationship between abstract mathematical structures and concrete particulars, the idea of an all-encompassing "environmental structure" that strings together independent projects also embedded the logics of operating a large-scale research project at the University of California, Berkeley and, as Alexander would soon consider, organizing a design consultancy.

Alexander discussed extensively the prospect of setting up a "company . . . to carry out conceptual design work" in a series of letters with architect Gerald Davis of Albert A. Hoover & Associates exchanged between March 1965 and March 1966.[112] Right after Alexander's departure from Berkeley, Davis wrote to him with a proposal to sell consulting services to a "real-life client." This was a medical education facilities programmer by the name of Lester Gorsline, who had the funds and eagerness to expand his understanding of architectural programming beyond traditional methods of balloon diagrams and schematic layouts.[113] Together with Berkeley faculty member Sim Van der Ryn, Davis was planning to use some of Alexander's methods and programs to perform the conceptual design and "form programming" for a new large-scale (upward of $30 million in budget) project by one of the universities for which Gorsline was working.[114] Davis and Van der Ryn agreed that before starting the project they ought to consult with Alexander about ways to implement his theories and methods in an applied project. Davis proposed to give Alexander compensation and credit for the use of his "creative thinking, and the programs" and potentially had his name featured in the company's masthead.[115]

Alexander was interested, but under conditions. For the company to work, he cautioned, it was essential that its information would be shared freely with everyone. Ideas, he wrote, were "universal," and for the company to restrict the use of methods, or computer programs, or information would be "entirely against the spirit of the whole venture."[116] The company's endeavor would be to, gradually, completely depict urban environmental structure.[117] "Each new project undertaken by a partner," he continued, "will deal with some known and unsatisfactory complex or set of complexes, and will reassort these complexes, reorganize them, and redefine their relation to all the remaining complexes in the city. <u>Only a company set up the way I am describing it, could undertake this</u> [underlined in the original]."[118] In order to achieve this tremendous pursuit that Alexander, in subsequent correspondence, estimated to require 40 years, the company also needed a specific governance structure. It would have to do away with "paternalistic company 'heads'" hindering the partners' independent and individual efforts. Alexander wrote:

> This company must, literally, belong to all its partners. For this reason it can of course not possibly be named after one, or two, or

three individuals, as you suggest. This is nothing but a hangover from the architectural firm, started by architects who love the sound of their own names more than the work they intend to do. This company needs a simple name, which says what the product is. The name I have in mind is Environmental Form Incorporated [underlined in the original].[119]

Alexander continued to outline a tentative structure for the company, including a sales division, the responsibilities of the partners, and a board of director that he proposed to chair. This structure corresponded to the three main aspects of every individual project: soliciting clients, the specific project and its conceptual design, and the theoretical framework that links the individual project to knowledge about the overarching environmental structure.

In his response, Davis elaborated on the governance structure that Alexander proposed and suggested that the company should be set up as a nonprofit corporation, to avoid paying income tax on accumulated capital and to facilitate contributions from philanthropic organizations and universities.[120] A few letter exchanges later, he explained to Alexander that a nonprofit corporation was the only plausible choice for the conceptual design endeavor, which was a "basic research job of the biggest order," one difficult to sell to clients and requiring as much subsidy funding as possible.[121] He also voiced a distaste for the corporate undertones of "Environmental Form Incorporated." He counterproposed names that would imply "a group of the highest scientific approach" such as Environmental Form "Group," "Laboratory," "Center," "Institute" or "Organization."[122]

Alexander persisted: "To me, who has had 20 years in universities, and no business experience at all, the whole ring of free enterprise and business and profit (as a measure of success and useful theory) is rather appealing . . . I still prefer EFI."[123] Despite being originally ambivalent about the nonprofit corporation proposal, Alexander quickly became convinced. Davis explained that the RAND Corporation, for instance, which Alexander presented as a model for EFI, was also not-for-profit.[124]

A recurrent point of contention in the correspondence was Alexander's influence in the company as the "intellectual fountainhead," "a master-genius-leader person" whose students indulged ideas of "in-ness" and engaged in "hero-worship."[125] Such concerns may have been exacerbated by his demand that the company employ predominantly students who had been trained at the University of California, Berkeley because they would be the only ones capable of grasping the "foreign conception of form" that EFI was promoting, one distinct from architects' common understanding of form as a "namby-pamby aesthetic thing which has to do with materials and their effects on people."[126] Alexander consistently appeased such concerns, often arguing that the enthusiasm of his students spoke to the validity of the ideas and not to his personality.

In October 1965, there was already a tentative formulation on a new Group at Berkeley and its offices. The Group would consist of 10 to 25 or more members, with a large number of graduate students and assistants at professional work. It would also include consultant programmers, social scientists, and other specialists. Office support would be provided by a small clerical staff under the direction of the secretary/treasurer. The offices would include several rooms for collaborative work, some individual rooms for the five directors of the group (the chairman, the president, and three elected directors with one-year appointments), a typing pool, and a large reference library.[127]

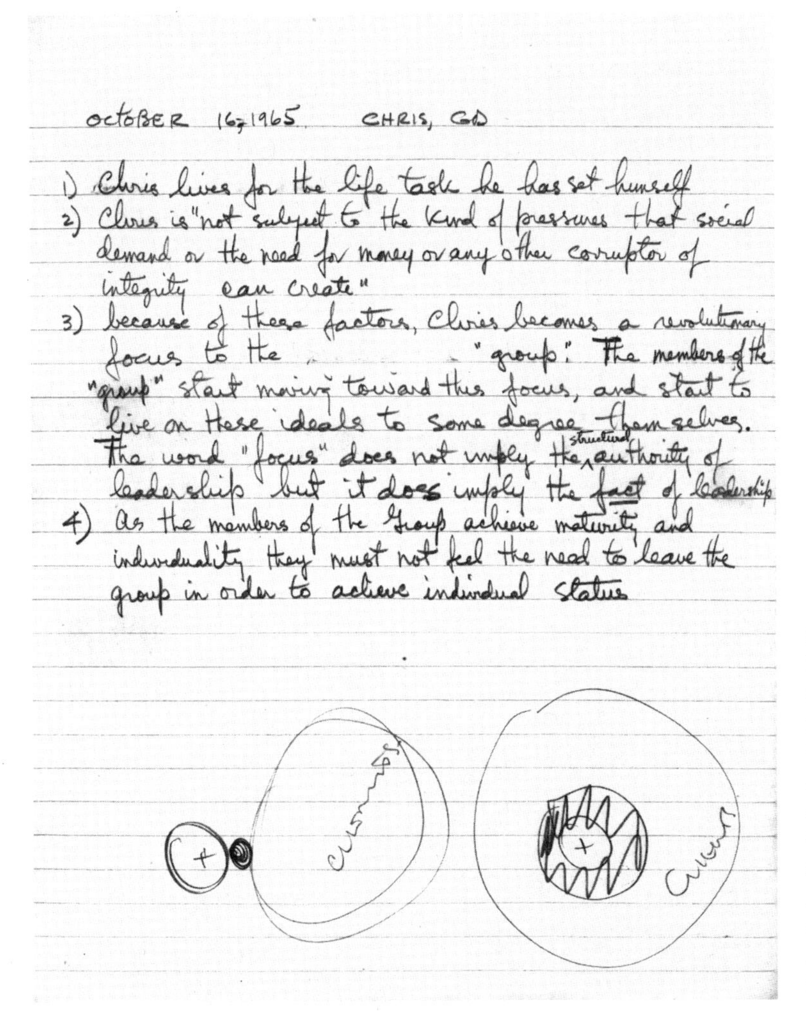
OCTOBER 16, 1965. CHRIS, GD

1) Chris lives for the life task he has set himself
2) Chris is "not subject to the kind of pressures that social demand or the need for money or any other corruptor of integrity can create."
3) because of these factors, Chris becomes a revolutionary focus to the "group." The members of the "group" start moving toward this focus, and start to live on these ideals to some degree themselves. The word "focus" does not imply the structural authority of leadership but it does imply the *fact* of leadership
4) As the members of the Group achieve maturity and individuality they must not feel the need to leave the group in order to achieve individual status

Figure 3.12
Davis's personal notes after meeting Alexander on 16 October 1965. Under the notes are diagrams thinking around the company's core and ways to achieve maturity and individuality. *Source*: Christopher Alexander / Center for Environmental Structure Archives, 1965.VIII.H.1_18. Courtesy of the Christopher Alexander / Center for Environmental Structure Archives.

Soon after his return from the UK, along with BART project collaborators Ishikawa and Silverstein and a starting grant from the Edgar J. Kaufmann Foundation, Alexander cofounded in March 1967 an independent nonprofit corporation that he named not EFI but instead Center for Environmental Structure. Until 1974, with the support of various research grants, the Center for Environmental Structure engaged in architectural and urban projects and produced several "pattern languages": diagrams for environmental "components" and a system of rules for combining them.

The baptism of the "pattern language" is said to have taken place in the summer of 1967, a few months after the establishment of the Center for Environmental Structure.[128] During a one-day seminar in Inverness, California, with a small grant from the American Bureau of Standards, hosting members of the National Institute of Mental Health, interdisciplinary experts from the East Coast including Marvin Manheim, and University of California, Berkeley faculty members and soon-to-be Center for Environmental Structure staff Sim Van der Ryn and Roslyn Lindheim, it was decided that environmental rules should be recast as "patterns" and their relations as a "language."[129] "Patterns" were not unlike the "diagrams" of the *Notes* in constituting "building blocks" of a design. Unlike diagrams, however, which were only valid in the context of a specific design brief and contingent on a specific data collection endeavor, patterns were seen as universal and reusable: as capturing relations that recurred in the environment. The language, in turn, encompassed the total set of rules by which these patterns were to be combined.

In a 1970 commentary on the Center for Environmental Structure's contributions, Roger Montgomery highlighted this "surprising . . . carefully intended shift . . . almost reversal" between the decomposition of a problem into a program, as presented in *Notes*, and the "combinatorial problems" of putting "pre-designed component images" together.[130] More simply put, the *Notes* gave the designer a process, a "program," for developing a design. This process took the form of a hierarchical tree. The designer had to "climb" the tree by producing unique diagrams that addressed the simpler groups of requirements hanging from each of the tree's branches and then combining the diagrams in the order indicated by the tree. In relational complexes, the designer could not see the mathematical structure of the process anymore. Instead, the designer was given the diagrams—now recast as "relations" or "patterns." Each pattern gave the designer some options for which pattern to use next. These links between the patterns were defined by the overall mathematical structure of the language that Alexander and his collaborators carefully authored. But the overall structure of the language was not disclosed to the designer using it.

Instead, the designer navigated that structure by following the patterns' local rules of combination.

In a tree structure, there could only be one way of moving from one diagram or pattern to the next. However, in pattern languages each pattern was linked to many other patterns. Instead of trees, pattern languages were structured based on "semilattices"—mathematical entities that still maintained a degree of hierarchy but allowed for multiple connections between each level of the hierarchy. Alexander had announced this move away from the hierarchical tree in one of his most famous articles that he titled "A City Is Not a Tree."[131] The article was published in a 1965 issue of the *Architectural Forum* and earned him the Kaufmann International Design Award, alongside renowned architecture critic Ada Louise Huxtable and architectural historian Lewis Mumford. In the article, Alexander argued for a more nuanced image of environmental complexity that could not be contained in the simplistic hierarchy of the tree.

Although Alexander promoted the move from the tree to the semilattice as a predominantly *theoretical* argument against the mathematical oversimplification of urban life—thus setting the stage for the article's subsequent reception—the move away from the tree was stimulated by *technical* concerns that had emerged from his work with HIDECS 2. As he wrote in one of his reports, when using a tree decomposition the computer program exhibited "irritating anomalies" and ended up separating requirements that were actually strongly connected.[132] In June 1963 he published a revision of the program, the so-called HIDECS 3, which addressed some of the issues by replacing the tree structure with a lattice structure in which requirements could be connected by more than one link. Along with the move away from the all-encompassing structure of the tree and the idea that sprang out of the BART project about the potential reusability of diagrams came the emphasis on *rules* contained within the diagrams or patterns.

After the BART study, Alexander and his Center for Environmental Structure collaborators worked on a series of consultancies where they advanced and tested the identification, combination, communication, and application of patterns. One of these projects was the development of a pattern language for multi-service centers—community centers offering social services to underprivileged groups—in Hunts Point, New York City, in the winter of 1967.[133] The Hunts Point multi-service center was developed by Alexander, Ishikawa, and Silverstein. Sandy Hirshen and Sim Van der Ryn also participated, along with the New York firm Gruzen & Partners. The project was in collaboration with Urban America's director and San Francisco Poverty Program participant Kenneth Simmons and was overseen by the New York City Human Resources Administration

(HRA).[134] The HRA's desire to develop a general planning scheme for multi-service centers promoted the prospect of the patterns' *reusability*—an idea that had already occurred to Center for Environmental Structure members after the highly laborious pattern development for the BART stations. Ishikawa and Silverstein developed 64 patterns and connected them in a mathematical structure that they called a "cascade."[135]

The "cascade" was a new take on what Alexander had previously called a "program" and was represented as a directed graph. The graph's nodes were patterns, and its links identified the order by which the patterns should be considered. The cascade was not to be confused with the language; it was merely a path through the language's structure. Alexander, Ishikawa, and Silverstein cautioned:

> Let us establish one thing at the outset. The language, and the cascade, are two different things. *The language contains far more structure than is captured in the cascade* [emphasis in the original]; the cascade is merely a partial representation of the language. However, we shall not discuss the additional structure in this report. Here, we confine ourselves, entirely, to those features of the language which are captured by the cascade.[136]

How to derive the language's overall structure was not discussed in the *Multi-Service Centers* report, yet the cascade had a somewhat hierarchical order, with bigger patterns considered first and smaller patterns later in the process. The report demonstrated the application of the language's patterns to design eight multi-service centers serving from 7,000 to 70,000 people at Hunts Point, San Francisco, Brooklyn, the Bowery, Phoenix, Newark, and Harlem, where the language was implemented to design two interconnected facilities.

The patterns themselves were rendered in an "if-then" format—where "if" corresponded to a (primarily social) situation, and "then" to a physical configuration adequate for that situation. The content of the patterns varied. Pattern number 4, for example, titled "community territory," mandated: "the service center is divided into two distinct zones, services and community territory; community territory includes space for community projects and a public arena."[137] The pattern directly engaged class and race—from the black power messages and civil rights movement figures in its opening collage, to critiques of multi-service centers' failures to fight poverty. Other patterns were focused on functional aspects of the centers, such as short corridors (pattern 31) and elevators/ramps (pattern 44).

An evolution of the *Atoms* ideas, patterns implicitly followed the motto: "A good environment is one in which no two tendencies conflict."[141]

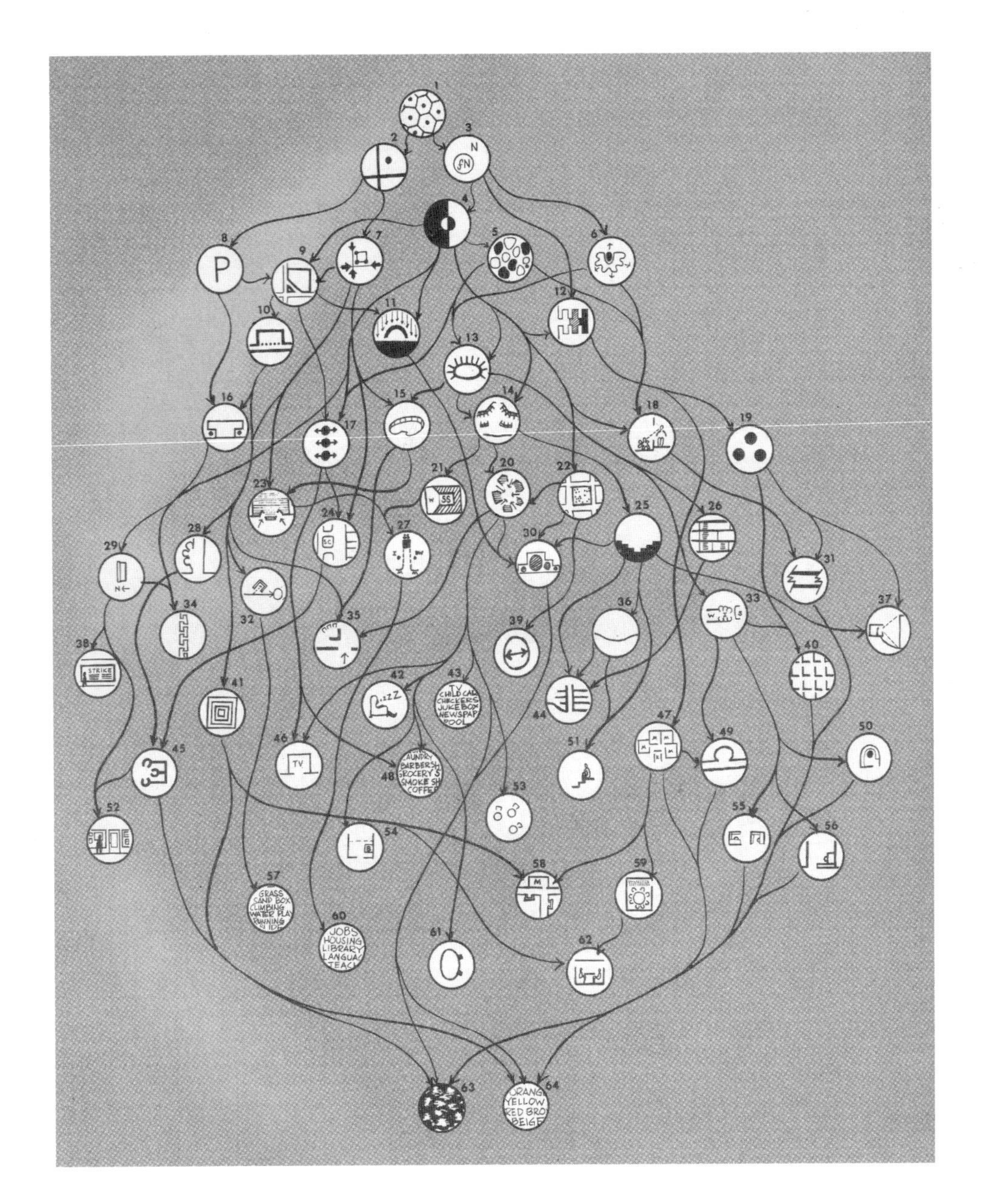

Figure 3.13
A cascade: a graph representing the structure and order of patterns. *Source*: Christopher Alexander, Sara Ishikawa, and Murray Silverstein, *A Pattern Language Which Generates Multi-Service Centers* (Berkeley, CA: Center for Environmental Structure, 1968), 18. Courtesy of the Christopher Alexander / Center for Environmental Structure Archives.

a

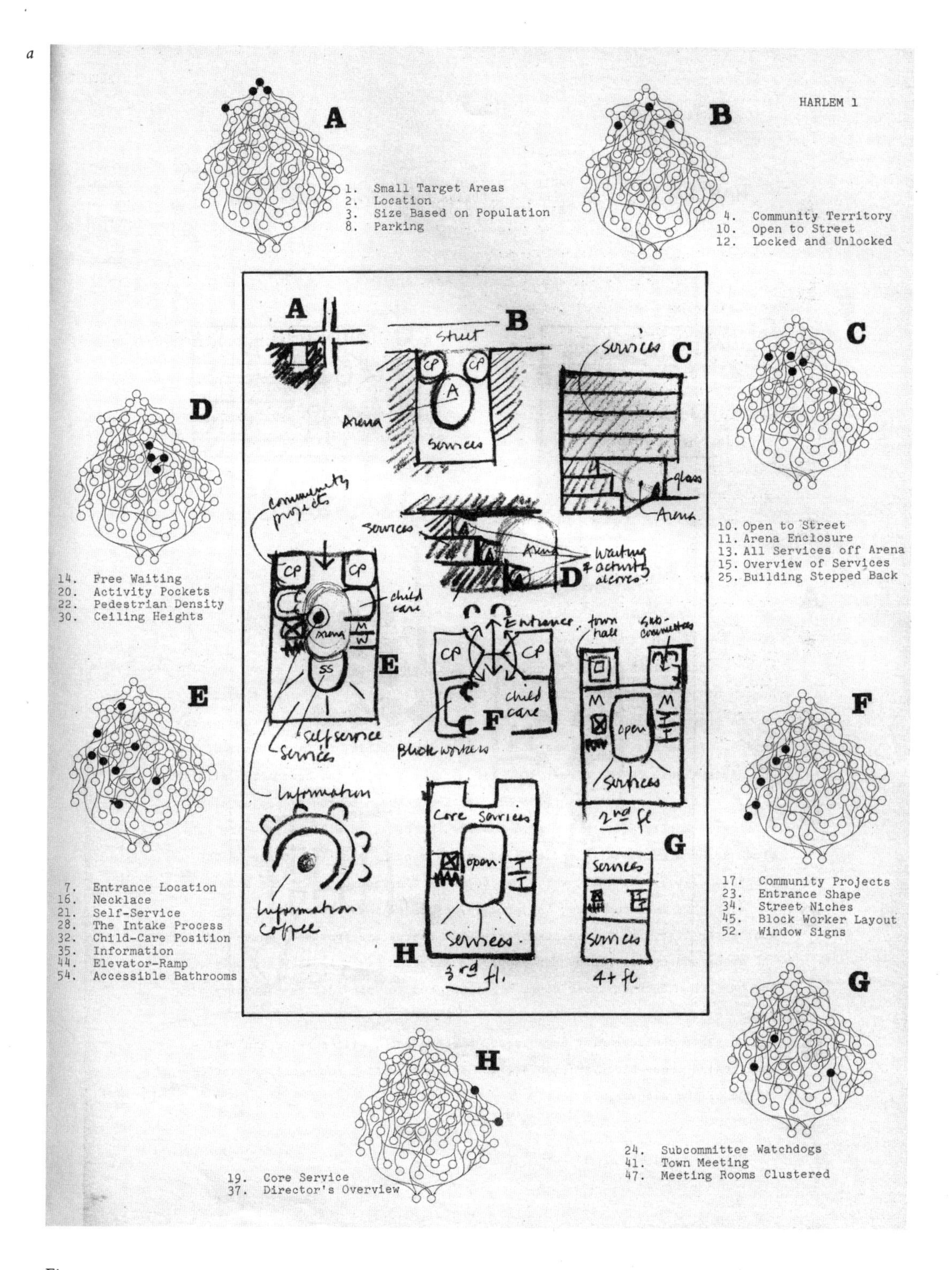

Figure 3.14
Using the cascade to design a multi-service center in Harlem. The black dots on the left are the patterns combined in the corresponding diagrams shown in the middle rectangle. *Source*: Christopher Alexander, Sara Ishikawa, and Murray Silverstein, *A Pattern Language Which Generates Multi-Service Centers* (Berkeley, CA: Center for Environmental Structure, 1968), 47 (*a*), 45 (*b*). Courtesy of the Christopher Alexander / Center for Environmental Structure Archives.

b

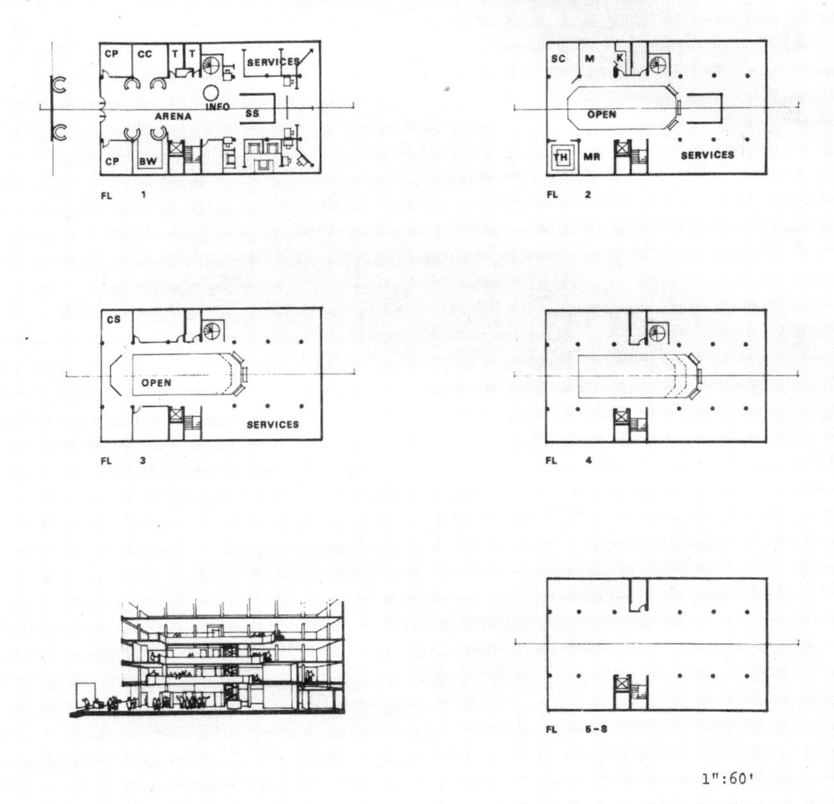
HARLEM 1
45,000 PERSONS - HIGH RISE - 12 SERVICES -
A "MOTHER" CENTER FOR TWO SMALL SATELLITES
(SEE HARLEM 2) - MULTI-LEVEL ARENA.
CP
CC
T
T
SERVICES
INFO
ARENA
SS
CP
BW
FL 1
SC
M
K
OPEN
TH
MR
SERVICES
FL 2
CS
OPEN
SERVICES
FL 3
FL 4
FL 5-8
1":60'
-45-

Interlude 6

As the anecdote goes, somewhere between 1967 and 1974, beat poet Allen Ginsberg performed a musicalized version of William Blake's 1794 poem "The Sick Rose" in front of a large crowd at the Berkeley campus central plaza.[138]

The spiritual take on Blake's pastoral elegy resonated with the late 1960s Berkeley climate of uprising against technocracy and the alienating forces of modernity. Not with dissimilar connotations, the poem also figured as the lyrical analogue for *A Pattern Language*.

Featured under an introductory section of the book titled "The Poetry of the Language," "The Sick Rose" provided not only an exemplar of *A Pattern Language*'s professed poetic potential but also a symbol of its social vision. Alexander and his collaborators aspired to reconstruct society by operating on they saw as its mirror structure, the built environment.[139]

Permeated by a "neo-romantic, community-anarchist, structuralist vision for a human city," as Silverstein would describe it, the book sought to restore within urban environments a kind of serenity reminiscent of the one that Blake portrayed in his *Songs of Innocence* and whose loss he lamented in his *Songs of Experience*.[140]

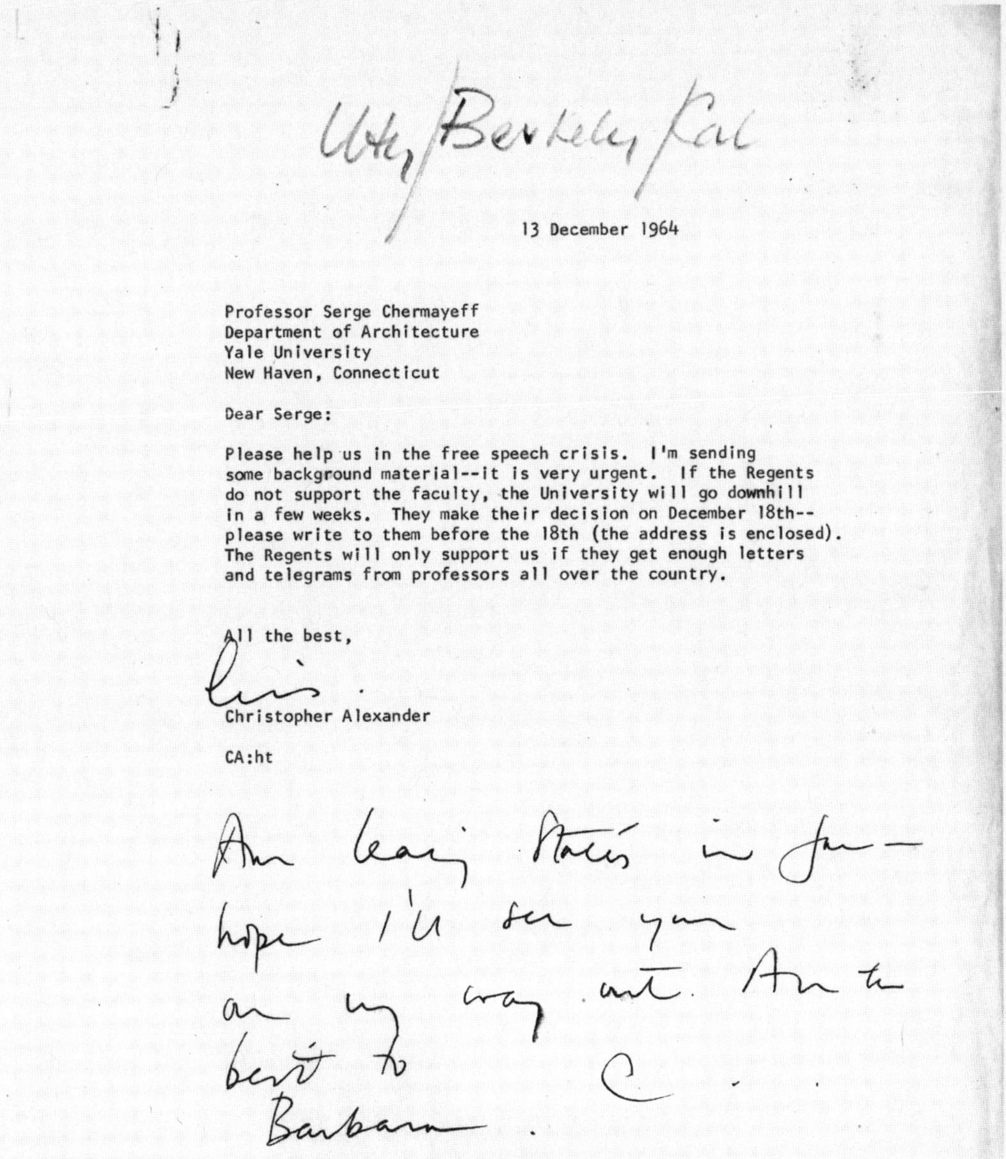

Univ, Berkeley, Cal

13 December 1964

Professor Serge Chermayeff
Department of Architecture
Yale University
New Haven, Connecticut

Dear Serge:

Please help us in the free speech crisis. I'm sending some background material--it is very urgent. If the Regents do not support the faculty, the University will go downhill in a few weeks. They make their decision on December 18th--please write to them before the 18th (the address is enclosed). The Regents will only support us if they get enough letters and telegrams from professors all over the country.

All the best,

Chris.

Christopher Alexander

CA:ht

Am leaving States in Jan— hope I'll see you on my way out. All the best to Barbara. C.

Figure 3.15
Source: Christopher Alexander, letter to Serge Chermayeff about the Berkeley Free Speech Movement. Christopher Alexander / Center for Environmental Structure Archives 1964. VIII. H.1. Courtesy of the Christopher Alexander / Center for Environmental Structure Archives.

The elimination of conflict carried emancipatory aspirations: "Life can fulfill itself only when people's tendencies are running free. The environment should give free rein to people's tendencies; conflicts between people's tendencies must be eliminated."[142] Such statements resonated with the National Institute of Mental Health's concern with alleviating environmental stressors and instilling environmental tranquility. In the summer of 1968, the Institute funded the Center for Environmental Structure with a $300,000 grant to develop the pattern language and communicate it to the public. The grant was paramount for the pattern language development.[143] It boosted theoretical research and turned the Center for Environmental Structure back toward the more "basic" aspects of the pattern language research: the language's structure and the formal definition of the patterns. The National Institute of Mental Health funding also helped staff the Center for Environmental Structure with graduate students who performed a series of "experiments," both in between and during the Center's project commissions. These experiments addressed the communication of patterns to users (their format and their level of emphasis and prescription), their level of abstraction (how generic they should be), and their structure (how they should be combined).

A driving question that developed after the Center had secured the grant concerned the pattern language's "generativity." From a theoretical perspective, a discussion on the topic was put forward by Alexander's 1968 article "Systems Generating Systems."[144] In the article he distinguished between "a system as a whole" and a "generating system." The former was a way of representing a structural property of a thing—one that could only be described by considering relations among parts. The latter was "a kit of parts" along with their rules of combination, and, as Alexander claimed, was necessary in order to produce holistic systems.[145] He wrote: "If we wish to make things which function as 'wholes' we shall have to invent generating systems to create them."[146]

From a practical perspective, the question about the language's generativity had emerged from efforts to use the pattern language as a design tool in Center for Environmental Structure design projects and academic seminars by Center for Environmental Structure staff at Berkeley and elsewhere. At Berkeley, Alexander ran several courses on the development and use of patterns. Ron Walkey tested the multi-service patterns with students from Berkeley's Department of Architecture. Silverstein at the University of Washington and Ishikawa at Berkeley also tested patterns in an educational setting.[147] These efforts indicated that the early versions of the pattern language could not produce designs only by following the language's structure. Designers needed to bring in professional expertise in order to go from

a

The W-algorithm.

1. Assume that every pattern locates at least one category.

2. Assume that categories are to be laid down in small sets, not one at a time - by a kind of yin-yang principle, which says, in effect, that categories X and Y must be laid down together, if there is a relevant pattern which locates both of them.

3. Now, a possible algorithm is the following.

Start with a category X.

Find all the brothersof X; and all the step-brothers of X.

Write these categories down as a set of sets. (Call this the W cluster)

Now write down all those patterns which locate any of these categories, in a list together.

Now, on scratch paper, make a list of all the categories which are generated out of any of these categories. (i.e. the categories which get located by patterns below the W cluster).

Take these categories one at a time (in some prescribed order), and form the cluster for it, and list its locating patterns.

There is no repeating of clusters, since this would be redundant.

This concludes the algorithm.

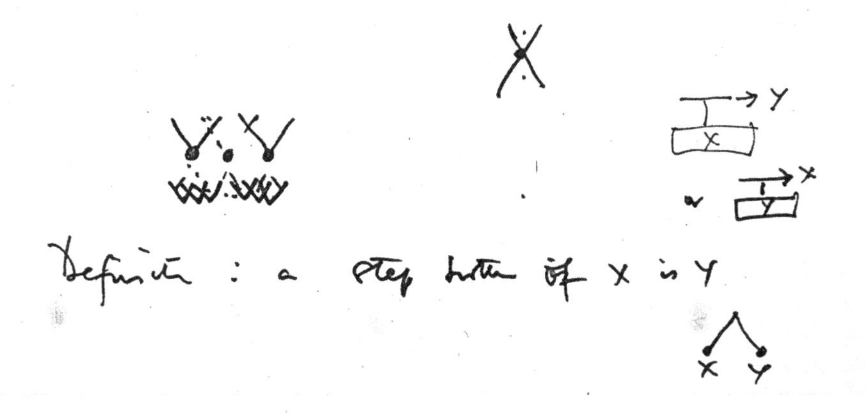

Figure 3.16 (*above and overleaf*)
Explorations on structuring the patterns. The "W-algorithm" and a list of categories. *Source*: Christopher Alexander/ Center for Environmental Structure Archives [1969.B.3]. Courtesy of the Christopher Alexander/Center for Environmental Structure Archives.

b

ACTIVE PLAY
ACTIVITY NUCLEUS
ADMINISTRATION
ALCOVE
ARENA EDGE
BAMBOO FOAM
BACKDOOR
BARNYARD
BATHING SINK
BATHROOM
BATHROOM FOR CHILDCARE
BEDROOMS
BEDS
BOOK STACKS
BUFFER ZONE
BUILDING
CARRELS
CBD
CEILING
CHILDCARE
CHILDCARE IN PUBLIC BUILDINGS
CHILD CAVES
CHILDREN'S BEDS
CHILDRENS CONSTRUCTION
CLASSROOM
CLOTHES DRYING CLOSET
COLORS
COMMUNITY
COMMUNITY CENTER
COMMUNITY FACILITIES
COMMUNITY PROJECTS
COMMUNITY TERRITORY
COMMUNITY WALL
CONCRETE BLOCKS
CORE SERVICES
CORRIDORS
COUNTER SURFACE
CUT RATE SPACE
DESK
DIRECTORS OFFICE
DRESSING
EARTH BERM
EATING
ELECTRICAL STRIP
ELECTRICAL SYSTEM
ELEVATOR
ENTRANCES
ENTRANCE SPACE
EVENING ESTABLISHMENTS
FAMILY ROOM
FLOORS
FLOWERS
FOOTBALL
FOUNDATION
FREESTANDING HOUSE
GARDEN
GATEWAY
GOSSIP SHOP
GYMNASIUM
HARD GROUND
HILL
HOUSE
HOUSE ENTRANCE
HOUSE LOT
HOUSING
INFORMATION
INTAKE
INTERVIEW
KITCHEN
KITCHENETTE
KITCHEN COUNTER
KITCHEN TABLE
LIBRARY
LIGHTS
LIVING AREAS
LOBBY
LOCAL FORUM
LOCAL OUTDOOR SPACE
LOCAL ROADS
LOCAL STORES
LOW INCOME COMMUNITY
MAJOR ROADS
MARKET
MASS HOUSING
MASTER BEDROOM
MEETING ROOM
MINI SCHOOLS
MSC
MUDROOM
NEIGHBORHOOD
NEIGHBORHOOD PRESCHOOL
OFFICE AREA
OFFICES
OLD PEOPLE
OLD PERSON'S DWELLING
* OUTDOORS
OUTDOOR PLAY
OUTDOOR ROOM
OUTDOOR SEATS
OUTDOOR STORAGE
OUTREACH
PANTRY
PARK
PARK BOUNDRY
PARKING
PARTITIONS
PASEO
PASSIVE HEAT
PATH EDGE
PATIO
PATIO ROOF
PAVING STONES
PEDESTRIAN PATHS
PLAY AREAS
PLAYROOM
PLUMBING
PLUMBING ACCUMULATOR
POCKET
POOL
PUBLIC BATHROOM
PUBLIC BUILDING
PUBLIC BUILDING ENTRANCE
PUBLIC GATHERING PLACE
PUBLIC OUTDOOR SPACE
PUBLIC SERVICES
RADIO?TV
RAMP
RECEPTION
REGION
REGIONAL FORUM
RESEARCH
RESEARCH OFFICES
ROADS
ROAD CROSSING
ROOFS
ROOM
RURAL
SALA
SAND PIT
SCHOOL
SCHOOL RESOURCE CENTER
SEAT
SECRETARY
SELF SERVICE
SHELVES
SHOWER
SIGNS
SINKS
SLEEPING
SOFT GROUND
STAFF LOUNGE
STAIRS
STORAGE
SUBCOMMITTEES
SUBCULTURE
SUBCULTURE BOUNDARY
SULPHUR
TABLE GAMES
TABLES
TOILETS
TOWN MEETING
TRANSITION SPACE
TREES
TV
URBAN AREA
YOLANDA

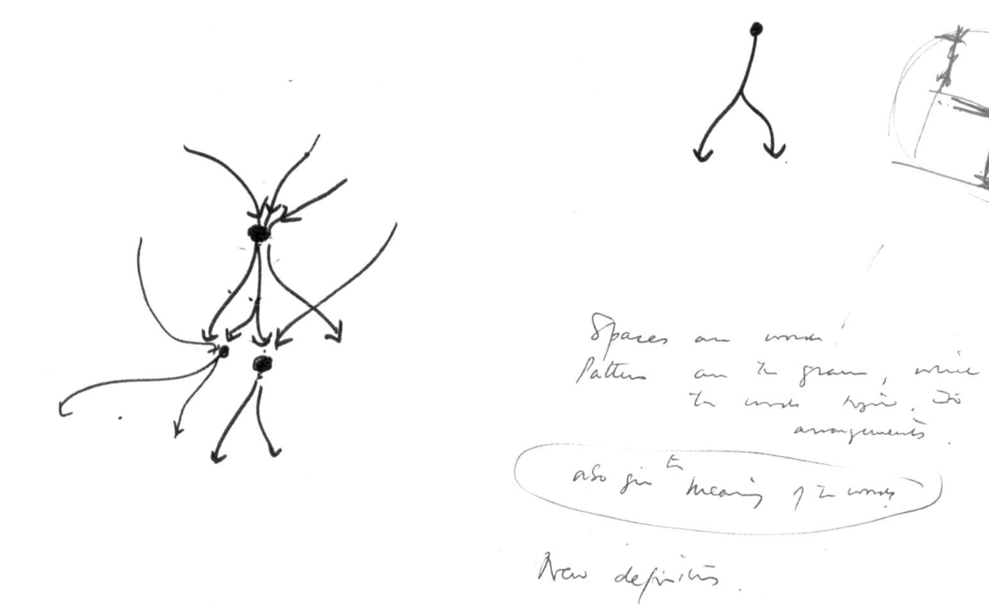

Figure 3.17
Sketches on the "W-algorithm."
Source: Christopher Alexander/Center for Environmental Structure Archives [1969.B.3]. Courtesy of the Christopher Alexander/Center for Environmental Structure Archives.`

the cascade to a schematic architectural design.[148] Alexander saw the language's generative capacity as being contingent on including the right "sequence" or "order" *within* the language's structure. Having nonarchitects successfully design with pattern languages was the next frontier. As we will see in the next chapter, the participatory aspects of *A Pattern Language* were as much a political as they were a technical project. If anyone could use the language successfully without prior architectural training, then the language could be said to be truly generative.

Generativities

From his time as a doctoral student in the late 1950s, Alexander's work was part abstract, part empirical. The relentless fieldwork and cataloging of requirements that he pursued in the early years of his PhD research was developed alongside logico-mathematical explorations of these requirements' "interlock." Variants of graphs, as abstract structures underlying the empirical world, appeared to accommodate this duality. In the *Notes* the tree ensured that, despite their immense empirical variety, requirements and their corresponding diagrams would speak the same mathematical language. Eventually, the "requirements" were transformed to "tendencies" and the "diagrams" to "relations" and then universal "patterns." The relationship between the ordering system behind the pattern language—a cascading network—and the empirical information contained in individual patterns followed a similar logic. The network persisted as a gatekeeper of the frictionless coexistence of architecture's physical receptacles and the life that they accommodated, mathematically enforcing an ideal of environmental tranquility. The network also ensured that each individual pattern describing a concrete relation between a physical setting and human behavior could be changed without compromising the stability of the overall language.

Unlike automatic layout algorithms where graphs functioned as *translation* devices by mapping and matching activity data with spatial relations in a one-to-one way, the pattern language's graph cascades operated as an *ordering* apparatus. They enabled calculating the ways in which individual patterns related to each other and the ways in which they were to be sequenced. They were the language's syntax—the underlying protocol for combining the patterns. By providing a mathematically correct structure for the language—as Alexander and his collaborators professed—graphs ensured that patterns were indeed generative: that, when combined, they still produced workable designs.

In automatic floor plan generation, graphs' generativity stemmed from their points' ability to represent activity-location dyads. It then sufficed to recombine the graphs' points, to connect them in different ways, in order to produce new spatial structures that could be evaluated for that given set of data. In pattern languages, graphs presented a contingent relationship with data. In the *Notes* it was interactions between the data attached to different design requirements that defined the specific form of the tree. Yet as multiple pattern languages started moving toward a singular pattern language, the network acquired a transcendental presence independent from and preceding the individual patterns.

Some of Alexander's critics saw the graph as what ethnomethodologist Michael Lynch has called "rhetorical mathematics": a visual representation giving the appearance of objectivity.[149] In discussing Alexander's work, for example, Geoffrey Broadbent pointed out that "highly personal views expressed in algebraic or graphic form have been given the semblance of objectivity by the medium of expression."[150] But the graph's rhetorical role was not only to legitimize Alexander's beliefs. Because of its visual presentation, the graph "pictured" the shape of Alexander's theories in a memorable and communicable manner and signposted their evolution—from the trees all the way to the networks. As a kind of "signature" or visual symbol, the graph was paramount for the dissemination and recognizability of Alexander's theoretical production. Complex formulas were hard to decipher and remember, but "trees," "semilattices," "cascades" spoke to the visually trained community of architects that Alexander was addressing. Visual rhetoric was also at work in automatic layout algorithms, where the visual similarity of graphs to string diagrams, or to various other ways in which architects had already been diagramming circulation, sanctioned graphs as transparent mediators between available data and algorithmically generated form.

The graphs' ability to serve as a model of empirical phenomena, which we saw in automatic layout work, was also avidly promoted by prominent mathematicians. Frank Harary, for instance, whom we earlier saw collaborating with Yona Friedman and Anthony Hill, coauthored a 1975 article provocatively titled "A City Is Not a Semilattice Either."[151] The title was of course a play on Alexander's "A City Is Not a Tree." The article acknowledged Alexander as the originator of graph theoretic analysis in architecture and urbanism, but lamented that, despite his self-issued mathematical corrective, his mathematics was still wrong.[152] The critique hinged on several points, the most salient being that Alexander used the graph as a tool for doing logical analysis as opposed to modeling empirically significant phenomena.[153]

Alexander did not exploit what Harary and his coauthor celebrated as one of this mathematical technique's most crucial utilities: its geometric (visual) presentation. To be sure, the different ways in which architects conceptualized and operationalized graphs as generative devices cannot be severed from cultures of abstraction. However, models—in our case mathematical models—do a kind of work that goes beyond representation and instrumentality. They work by making knowledge communities believe in the fictions that they help produce and follow them by reenacting their rules of abstraction.[154]

Despite rifts between automatic layouts and pattern languages, graphs worked through their mathematical properties and visual traits, to entrench the fiction that form could be reliably generated from data through rule-based, stepwise procedures immune to erroneous human interventions. Of course, human designers still placed doors and windows on automatically generated floor plans and gave geometric form to the patterns as they were navigating the language that contained them. However, steadfast mathematical skeletons determined the resolution of these concrete architectural particulars. This, after all, is the essence of graph vision: the prioritizing of abstract underlying structures over sense-perceptible appearances and the subjugation of possibility under these invariant structures' certainties.

Graphs' immaculate topologies obfuscated the messiness and transformative power of human judgments and decisions by taming them within algorithmic processes. Mobile while immutable, or sometimes precisely because they were mutable, these algorithms effaced the dependency of their inception on data—as threat, resource, and possibility in postwar architectural research.[155] Graphs' mutations continued as intellectual and political climates shifted at the end of the 1960s. Moving on to graphs as infrastructures, we will now delve into such relationships between graph-powered processes of enumerating combinations between discrete entities and ideals of open-endedness, choice, and change that took center stage in the theories and politics of participatory design in the 1970s.

4 Infrastructures

Control

Among some six hundred pages of the much-anticipated sequel to the iconic countercultural compendium *Whole Earth Catalog* was a full-page entry on *A Pattern Language*.[1] Featured in a section on "Soft Technology" titled "Everything Design," the entry, written by the author of the 1975 utopian novel *Ecotopia* Ernest Callenbach, lauded *A Pattern Language* as "the most important book in architecture and planning for many decades."[2] "No possible doubt," Callenbach wrote, "it's simply a great book—a bracing adventure for architectural thought, a lift for the spirit, an inspiration for practical work!"[3] Complementing this hearty praise was a one-line comment by the *Catalog*'s editor Stewart Brand: "*I suspect this is the best and most useful book in the* Catalog—SB."[4]

The book's exuberant embrace by one of the most influential publications on do-it-yourself culture and creative communalism asserted *A Pattern Language*'s position as a bible for community engagement and participatory design—a reputation that has followed it until today. The idea of "participation," in its many, and often muddled, methodological and conceptual expressions, animated architectural debates in the late 1960s fueled by political and social fermentations and growing disbelief in a fraught design professionalism.[5] In the United States, the civil rights movement and the 1968 Fair Housing Act amplified efforts to grant social groups affected by design and planning choices

Figure 4.1 (*opposite*)
Site plan and houses in the site, Proyecto Experimental de Vivienda, Lima, Peru. *Source*: Christopher Alexander, Sanford Hirschen, Sara Ishikawa, Christie Coffin, and Shlomo Angel, "Houses Generated by Patterns (Report on the Project Experimental de Vivienda)," 1969. Christopher Alexander / Center for Environmental Structure Archives, 1969.I.A.1.9 (*a*), 1969.I.A.1.10 (*b*). Courtesy of the Christopher Alexander/Center for Environmental Structure Archives.

agency and control over the decision-making process. The University of California, Berkeley, where Alexander taught, was an epicenter of such efforts. Political involvement across campus was swaying former acolytes of science and rationality toward criticizing the limits of information-based processes for reasoning about social phenomena. Horst Rittel, for instance, chief instigator of "design science," was publishing on the limits of information-based prediction in planning and developing systems for argumentative decision-making.[6]

The Center for Environmental Structure's work on pattern languages, boosted by the hefty patronage of the National Institute of Mental Health, served as fertile ground for applied and theoretical work in participatory design. Between 1969 and 1971 the Center embarked on two major projects of participatory architecture: the Proyecto Experimental de Vivienda in Peru in winter 1969 and the campus planning process for the University of Oregon campus in Eugene in 1971. The Proyecto Experimental was an experimental low-cost housing project of a total of 1,500 dwellings in Lima spearheaded by Peru's president Fernando Belaunde Terry, who was also an architect, in collaboration with the United Nations.[7] The Center for Environmental Structure was selected to participate in the competition together with 12 other international and 13 local architectural firms. The competition submission included a site plan, house plans, and a construction system, and required for each scheme to determine how it would grow and adapt with time. The project began with several members of the Center, including Alexander, living with families in Lima to document their lives and activities. These observations were then taken back to Berkeley to develop patterns in parallel with specific designs. The competition led to the realization of one cluster of houses accompanied by publication of the design guide *Houses Generated by Patterns*.[8] In her 1984 visit to the block designed by the Center for Environmental Structure, Ishikawa reported finding inhabitants "very happy and proud of their home and neighborhood" and having "made additions and modifications to their houses."[9]

In 1971, the Center also proposed a process for planning the campus at the University of Oregon with the University of Cambridge individual colleges' locations and relations as reference. This was published as the *Oregon Experiment*, in which the Center outlined principles for the growth of the campus over time through the participation of its occupants, including ideas such as organic order, participation, piecemeal growth, patterns, diagnosis, and coordination.[10] The use of the language by nonarchitects in the context of participatory projects proved a taxing challenge. It produced, as Silverstein lamented, "very sketchy, rambling designs which did have certain holistic properties,

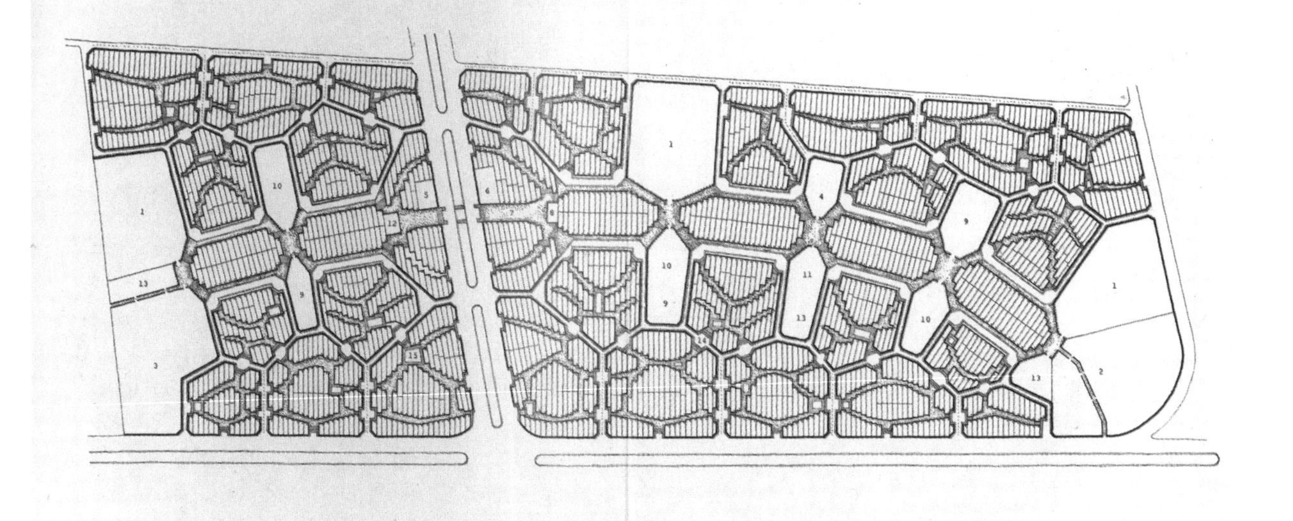

1 Primary School
2 Secondary School
3 Technical Secondary School
4 Church
5 Cinema
6 Supermarket
7 Market
8 Municipal Offices
9 Grove of Trees
10 Kindergarten
11 Clinic
12 Dance Hall
13 Sports Center
14 Parking
15 Outdoor Room

North

0 50 100 200 Meters

THE SITE

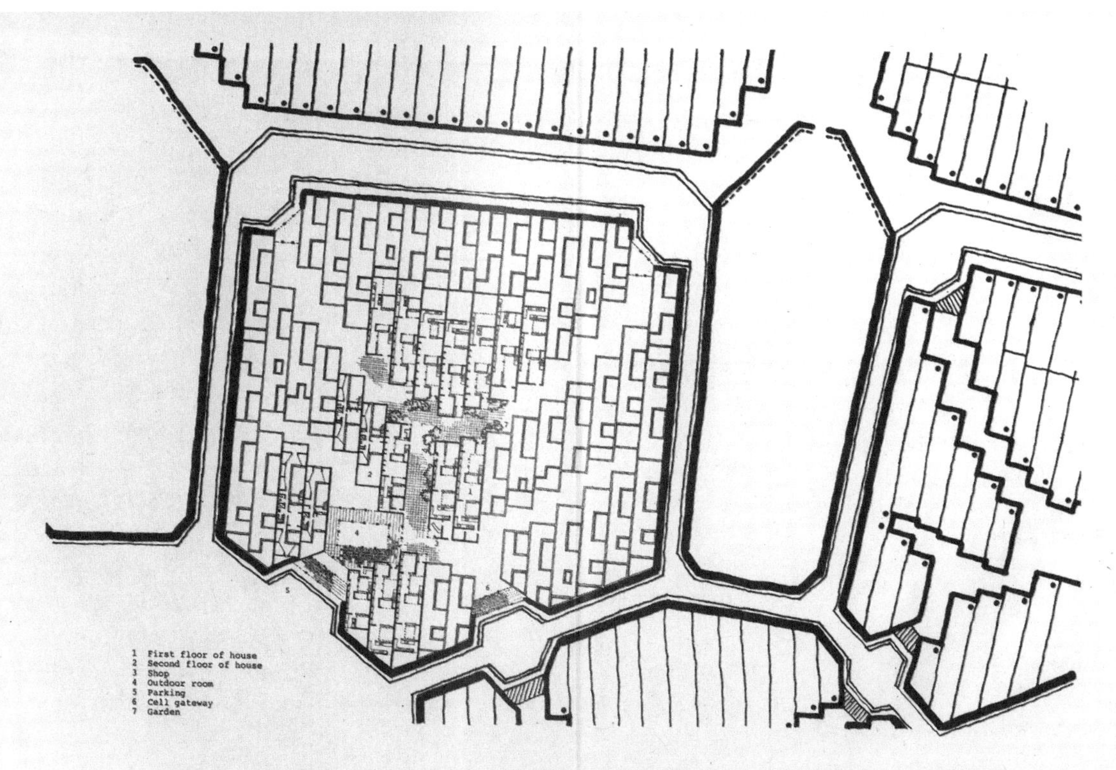

0 5 10 20 50 Meters

HOUSES IN THE SITE

but which were not yet clearly formed buildings."[11] This conundrum appeared to be caused by flaws in the sequencing of the patterns.

Participation was less a political imperative than testbedding—prime ground for advancing mathematical development of the language. For pattern languages to be *generative*, so that anyone, not only trained architects, could use them to design, they needed an accurate and systematic structure. Alexander told biographer Steven Grabow that in 1972 he enlisted his PhD students to describe the mathematical attributes of such a structure using the case of a simple language for generating Japanese teahouses consisting of 24 patterns. He was confident that the sequence of the Japanese teahouse pattern language was successful because its patterns were structured in such a way that when they were presented to someone, "that person would then form a complete image of a teahouse in their mind."[12] The sequence that worked in the way Alexander described existed within a gargantuan space of possible sequences; 24 factorial to be precise—620,448,401,733,239,439,360,000 possibilities of sequencing the teahouse patterns. Alexander gave his graduate students the problem of identifying what made the Japanese teahouse sequence of patterns work and whether there were other possible sequences that could work in a similar way. In reminiscing about the event, he asked: "You could tell intuitively that it was great; but if one had made a cascade of the type we were studying earlier, what would have been the mathematical properties that got from that cascade to the correct sequence?"[13]

The problem of the language's structure occupied much of Alexander's intellectual efforts while he headed the Center for Environmental Structure. The patterns could not work unless they sat on the right skeleton. Once the right skeleton had been discovered, it would become possible to write a computer program allowing anyone to automatically navigate from one pattern to the next while they were working on a design. Initially, as Alexander recounted to Grabow, *A Pattern Language* was in the form of a computer database with a series of computer programs extracting pattern sequences from the pattern cascade.[14] It turned out that a Xeroxable book was just as good.[15] The idea was to embed the mathematical structure underlying the patterns—the cascade—in the accessible format of a "cookbook": the do-it-yourself architecture manuals circulating in countercultural periodicals such as the *Whole Earth Catalog*.

The cascade was the informational structure that held all patterns together. The cascade and the patterns, combined, held the possibility of countless—or theoretically countable, but too many to practically achieve—individual expressions using it. Different paths through the language meant different designs. So did different concrete realizations of the diagrammatic patterns. Alexander, it would seem, had come full circle from the time when, while documenting life and architecture at Bavra, India, he had embarked on conversations with eminent Indian architect and environmental design pioneer Balkrishna Doshi. This long-standing conversation concerned, as Doshi put it, "the problem of shaping the environment in a very special light," confronted with India's material shortage, scant technical knowledge, and rapid growth.[16] The exchange resulted in a coauthored presentation in the 1962 installment of the famous International Design Conferences in Aspen.[17] An edited form of the presentation was published in the journal *Landscape* with the title "A Role for the Individual in City Planning: Main Structure Concept"—this was the article where we earlier heard Alexander and Doshi agonize over the displacement of architects by various experts.

To the threat of architects' agency being reduced to the beautification of an unwieldy technological environment, Alexander and Doshi proposed a remedial solution. Architects, they argued, could still contribute to the design of the environment in a way no specialist could.[18] This would be by specifying "the overall organization of the environment" while leaving it to the individual "to control and construct his [sic] immediate surroundings."[19] This is what Alexander and Doshi called "the main structure concept": the identification of a "structure" and "filler" at any scale of resolution, where the structure was a skeleton necessary for the filler to exist.[20]

The "structure-filler" idea was not particularly original. Alexander and Doshi credited the Japanese metabolists Kenzo Tange and Kiyonori Kikutake, but also Hong Kong's refugee housing that provided a mere six-story "artificial ground" for occupants to develop their own dwellings.[21] The structure-filler model was popular with groups such as Team 10 who advanced a form of urbanism where architecture provided a structure (a "stem," a "group," a "network") corresponding to different forms of social organizations and enabling a degree of adaptation to the inhabitants' needs.[22] Possibly aware of the conceptual affinities between his mathematical work and Team 10's concern with structure, Alexander presented his Indian village worked example at the September 1962 Team 10 meeting in Royaumont, where it received mixed reactions.[23] This was weeks before he found a more receptive audience at the first Conference on Design Methods at Imperial College, London. As we saw, the structure-infill scheme was also enabling bold proposals for mobile dwellings within ever-expanding spatial infrastructures from the members of the Groupe d'Études d'Architecture Mobile, led by Yona Friedman. Meanwhile in the Netherlands, architect and design theorist John Habraken was detailing groups of prefabricated elements building up "innumerable dwelling types"

Interlude 7

In July 1976, the audience of the 3rd Annual Conference on Computer Graphics, Interactive Techniques, and Image Processing was presented with screenshots of a computer-aided design system named YONA. Photostat reductions and reversals of 4-by-5 Polaroids, the screenshots showed graphs generated during the system's operation and visualized on a touch-sensitized IMLAC PDS-1D. Written in PL/1, YONA was running on the Interdata minicomputers that equipped a computing facility in the MIT Department of Architecture that went by the name the Architecture Machine.

One screenshot depicted five crosshair-drawn points, labeled with names of domestic architectural spaces, and connected with lines to shape a graph that whimsically looked like a child's drawing of a house. Another screenshot showed osculating b-spline curves fitted between the graph's dual and its offset boundary to form a "bubble diagram"—the architectural parlance for a rough sketch of a plan indicating the locations and interconnections of spatial enclosures. A third screenshot pictured a schematic drawing of an architectural plan, where the "bubbles" were rationalized into straight lines representing wall boundaries.

The screenshots visually summarized a graph-theory-based computer-aided design process for do-it-yourself designers. Sitting in front of the computer screen, the "unpracticed designer," as the conference paper called the system's user, started by selecting a desired architectural program (types and sizes of spaces) and their connections.[25] The user-specified spaces became the points ("nodes") of a graph, and the links became the lines. A planarity test checked whether a one-level arrangement of the architectural program was possible or whether the user needed to make modifications, and outputted an image of a labeled planar graph with a preliminary site positioning that the user could rearrange by moving nodes on the screen. Once a desirable arrangement was reached, the system presented the corresponding "bubble diagram" to the user, who could then shape it into rooms by snapping straight lines and curves on a grid projected on top of the "bubbles."

An homage to Yona Friedman who furnished the system with its "underlying philosophy" and an acronym encapsulating the system's aspirations, Your Own Native Architect or YONA strove "to allow people to design their own homes without either a middleman or a middle machine creating whole solutions for them."[26]

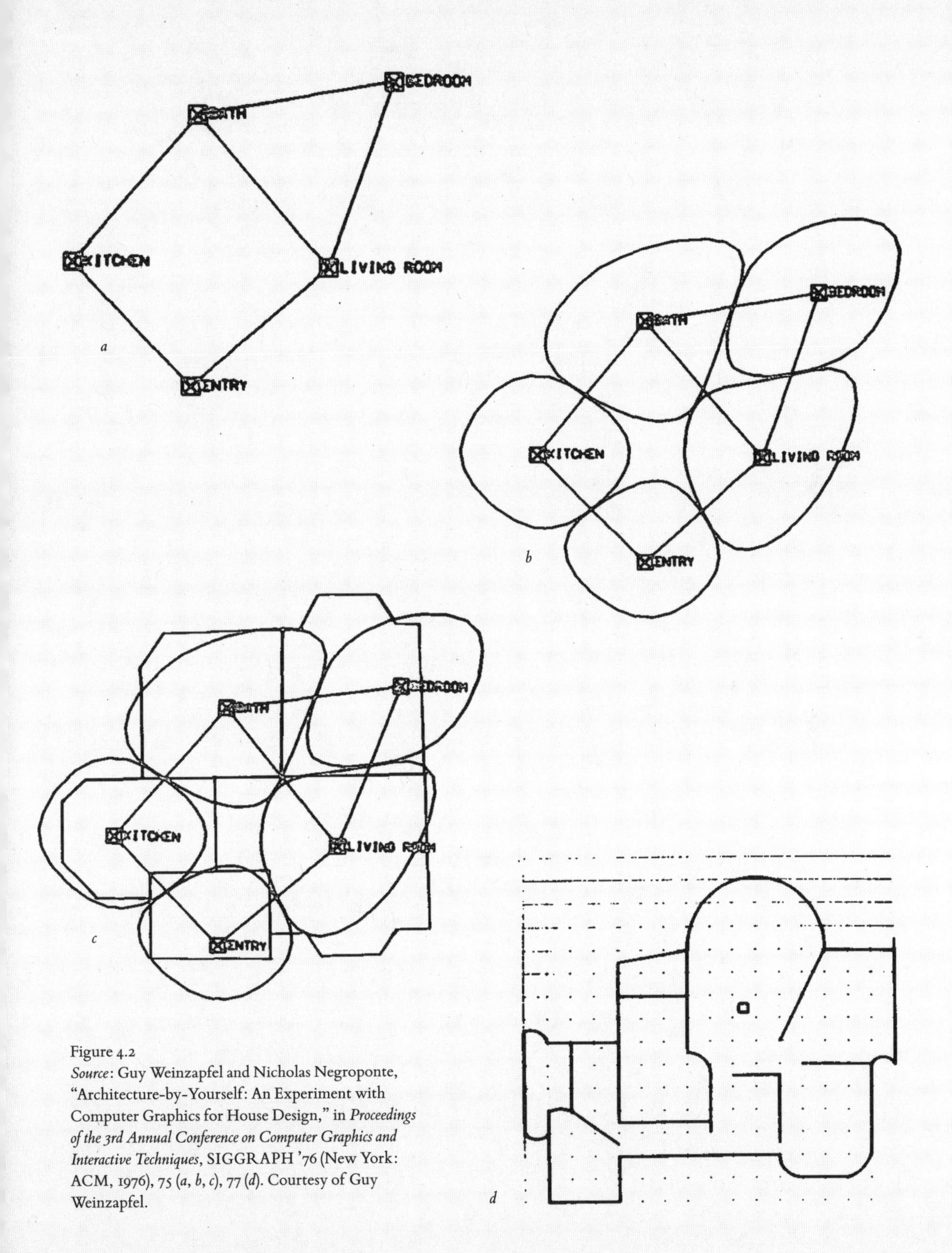

Figure 4.2
Source: Guy Weinzapfel and Nicholas Negroponte, "Architecture-by-Yourself: An Experiment with Computer Graphics for House Design," in *Proceedings of the 3rd Annual Conference on Computer Graphics and Interactive Techniques*, SIGGRAPH '76 (New York: ACM, 1976), 75 (*a*, *b*, *c*), 77 (*d*). Courtesy of Guy Weinzapfel.

to match individual preferences, supported by structures "growing, developing and changing."[24]

The structure-filler model—Alexander and Doshi's "main structure concept" which, recalling Friedman, I will refer to as the "infrastructure model"—was the staple of 1960s participatory design theories and architectural proposals. And yet, putting Alexander's 1962 paper with Doshi and *A Pattern Language* side by side reveals an important translation of these physical propositions' condition to the theoretical realm of a method for designing. *A Pattern Language* emulated the "main structure concept" of participation, but not in the form of a physical infrastructure. Instead, it put forward a *virtual* infrastructure that, not unlike its physical counterparts, delimited possibilities of choice and ensured that they all fell within the system used to generate them. The infrastructure, in the form of the cascade-graph, captured the unchanging, the essential, that which lay *beneath* all individual realizations.

The politics of *A Pattern Language* as a tool for individual and collective design empowerment lay in its all-encompassing skeleton and the tenacious ways it supported individual choice. The story of graphs as infrastructures for participation, which we will continue with now, is a story about the mathematical configuration of design possibility, the generation of ranges of alternatives, and the protocols for their navigation. It is also a story about graphs receding from view, as infrastructures do, to serve as the calculable underbelly of ambitious proclamations of freedom and individual expression. It is also, as always, a story of institutional infrastructures: the mobilizing of computing resources, interdisciplinary expertise, and grant funds to render the calculus of choice and possibility a robust and versatile research agenda.

Change

In 1970, one year before the publication of *Pour une architecture scientifique*, Yona Friedman was invited to participate in the 1970 Osaka Expo, a world's fair remembered for its emphasis on multimedia and information technologies. Responding to the Expo's theme "Progress and Harmony for Mankind," Friedman proposed a machine that he called the FLATWRITER as the implementation of the graph-based theory of scientific architecture that he had been developing since 1964.[27]

Friedman presented the FLATWRITER machine as an "application of the repertoire": the process of making "menus" of all possible architectural plans for a given number of rooms. Conceived as a design typewriter of sorts, the FLATWRITER presented the user with two keyboards, one consisting of abstract geometric shapes and

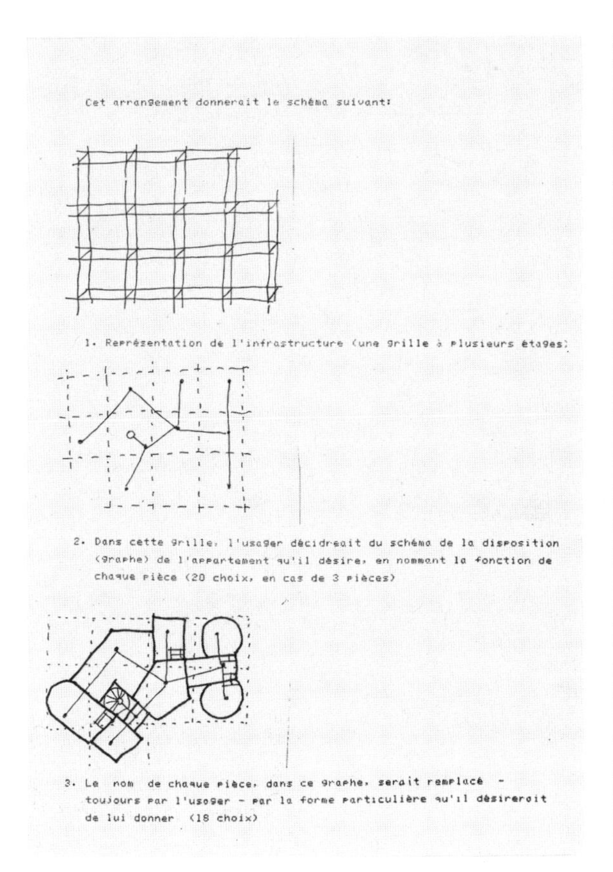

Cet arrangement donnerait le schéma suivant:

1. Représentation de l'infrastructure (une grille à plusieurs étages)

2. Dans cette grille, l'usager décidreait du schéma de la disposition (graphe) de l'appartement qu'il désire, en nommant la fonction de chaque pièce (20 choix, en cas de 3 pièces)

3. Le nom de chaque pièce, dans ce graphe, serait remplacé - toujours par l'usoger - par la forme particulière qu'il désireroit de lui donner (18 choix)

a

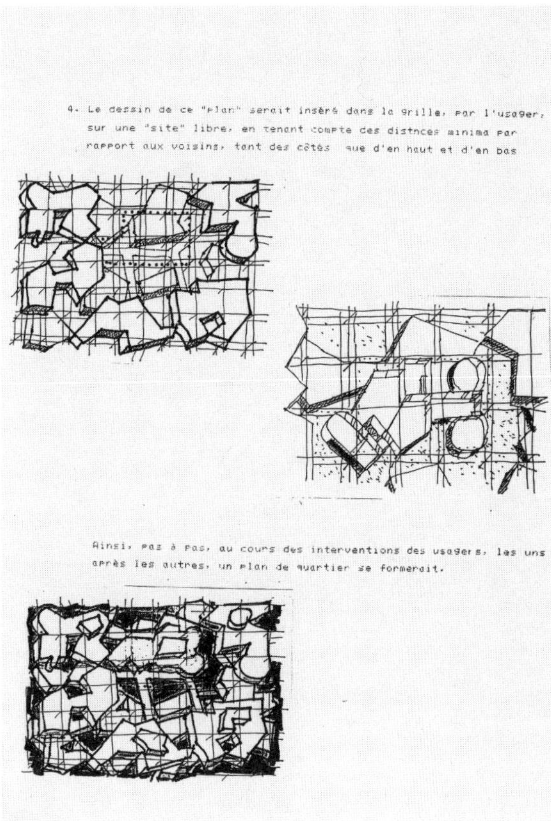

4. Le dessin de ce "plan" serait inséré dans la grille, par l'usager, sur une "site" libre, en tenant compte des distnces minima par rapport aux voisins, tant des côtés que d'en haut et d'en bas

Ainsi, pas à pas, au cours des interventions des usagers, les uns après les autres, un plan de quartier se formerait.

b

Figure 4.3
FLATWRITER instructions, typed text. *Source*: Fond Denise et Yona Friedman, Folder 300, pages 331 (*a*), 332 (*b*). Courtesy Fond Denise et Yona Friedman, Archives photographiques Jean-Baptiste Decavèle, Marianne Friedman Polonsky, https://www.yonafriedman.org/.

equipment types that assigned function to the shape, and one consisting of "weights" denoting the frequency at which the system's user—in this case the future inhabitant of the dwelling designed by the machine, whom Friedman referred to as the "future user" or "Mr. X"—visited each space type in their everyday life.

The machine, Friedman observed, "predicated a framework of existing stocks of prefabricated elements, service units, bathroom and kitchen units."[28] With these keyboards, the user could "type" and "print out" preferred plan configurations within a "repertory of several million plans," while the machine fed back "warnings concerning the consequences implied by any projected use pattern."[29] The FLATWRITER calculated such "warnings" on the suitability of each floor plan configuration for a given set of activities. In *Pour une architecture scientifique* Friedman likened this process to eating out, choosing from a restaurant menu through a compromise of preference and price.[30] The FLATWRITER posed as a device for achieving nothing less than architecture's "democratization."[31]

a

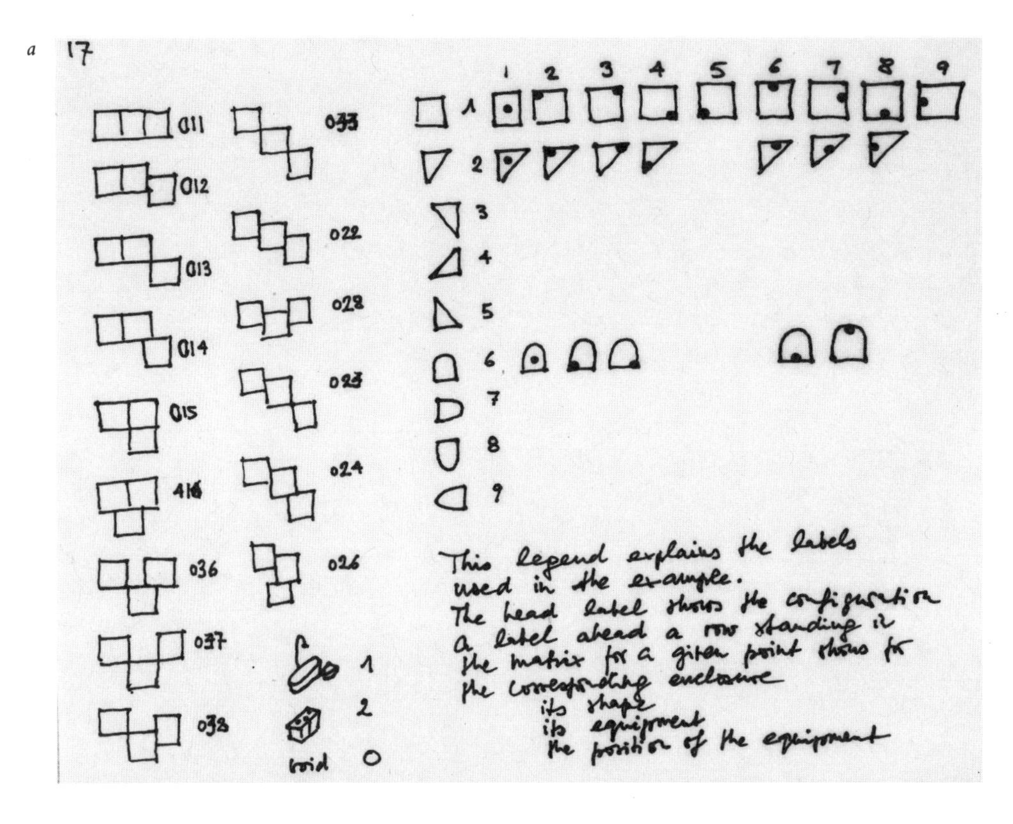
17
011
012
013
014
015
036
037
038
033
022
028
023
024
026
1 2 3 4 5 6 7 8 9
1
2
void 0
This legend explains the labels used in the example.
The head label shows the configuration
a label ahead a row standing in the matrix for a given point shows for the corresponding enclosure
its shape
its equipment
the position of the equipment

b

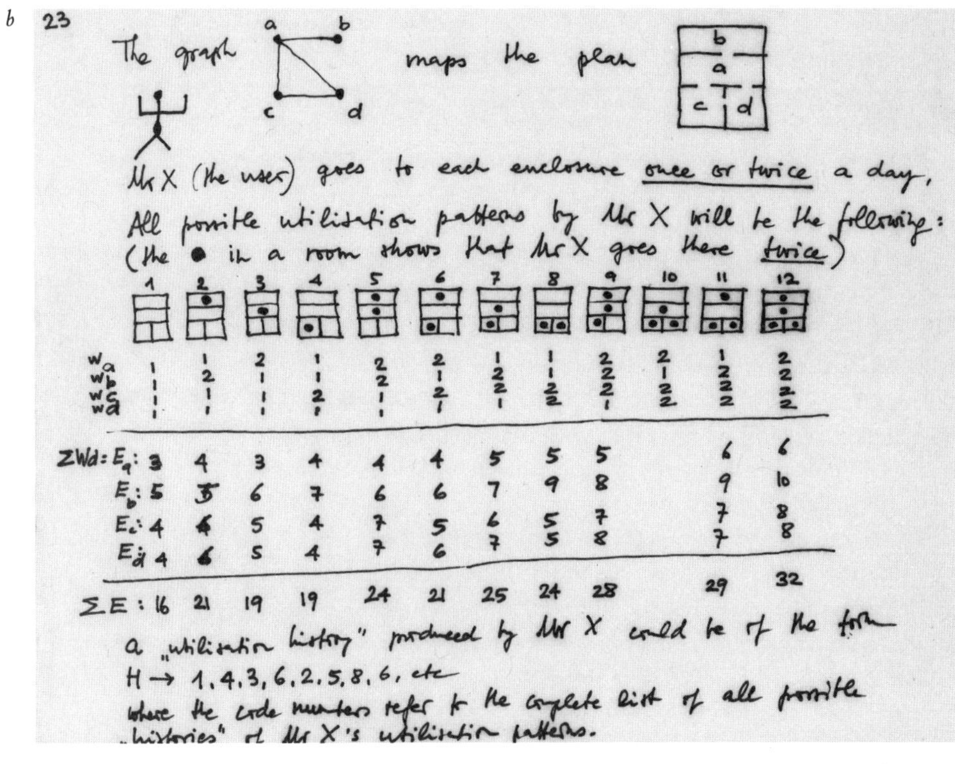
23
The graph
a b c d
maps the plan
Mr X (the user) goes to each enclosure once or twice a day,
All possible utilisation patterns by Mr X will be the following:
(the ● in a room shows that Mr X goes there twice)
ΣE : 16 21 19 19 24 21 25 24 28 29 32
a „utilisation history" produced by Mr X could be of the form
H → 1, 4, 3, 6, 2, 5, 8, 6, etc
where the code numbers refer to the complete list of all possible „histories" of Mr X's utilisation patterns.

Central to the process was the idea of informed decisions—through "warnings"—in the form of an efficiency metric Friedman called "effort."[32] The concept came from his work on "urban mechanisms"—systematic ways of describing and intervening in the urban settlements potentially developed within an "ideal infrastructure."[33] The argument for an "ideal infrastructure" first appeared in a transcript of a lecture that Friedman delivered at Harvard University and at the Carnegie Institute of Technology in March 1964. It was published in *Architectural Design* the same year under the title "Toward a Coherent System of Planning," next to an order form for Alexander and Chermayeff's just circulated *Community and Privacy*. In these lectures, Friedman endeavored to establish a set theoretic formulation for enumerating social organizations and argued for a city planning model—the "ideal infrastructure" model—that would permit seamless transitions between distinct social structures.

"Urban mechanisms" were the operating principles of the infrastructure. In a 1967 article in *Architectural Design* where he developed the concept, Friedman outlined an informational process where inhabitants' activities are taken as "input" and used to calculate configurations of elementary volumes as "output."[34] By now we should be familiar with such notions from work on space allocation and automatic floor plan generation, which, we have seen, drew from industrial engineering methods. Friedman attempted to mathematically describe not only the static three-dimensional infrastructure of the Ville Spatiale but also the mobile fillings within the infrastructure as a network of departure and arrival points labeled with different functions.[35] Observing the frequency of movements between departure and arrival points would allow mapping the behavior pattern of the city's inhabitants.[36] The efficiency of a specific urban configuration for conducting these activities constituted the "urban mechanism's" "effort."[37]

To model an "urban mechanism," Friedman wrote, one needed "a list of possible configurations of a set of obstacles in a limited field; a list of possible distributions of frequencies of movements between couples of such obstacles (here meaning departure points and targets); a calculation of the total overall efforts deployed by the inhabitants (individuals or groups) in their movements."[38] A computer, he suggested, could enumerate the "full range of possibilities" of urban configurations along with corresponding "warnings" for the efficiency, "effort," for its inhabitants.[39] Apart from a "warning device for the planner, urbanist, and sociologist," Friedman's system would also serve as a "research tool" that would keep track of different activity patterns and physical configurations, possibly revealing underlying patterns.[40] These ideas were carried over into the FLATWRITER machine but adapted, in its so-called "first loop," as a self-tracking and self-optimization process.

Figure 4.4 (*opposite*)
The future user: manual on architecture (n.d., circa late 1960). Draft sketches on the operation of the FLATWRITER. Keyboard indicating position, shape, equipment placement, and site orientation, accompanied with a sample sequence of keystrokes (*a*), and "effort" calculations based on the daily habits of "Mr. X" (*b*). *Source*: Fond Denise et Yona Friedman, Folder 293, pages 230 (*a*), 240 (*b*). Courtesy Fond Denise et Yona Friedman, Archives photographiques Jean-Baptiste Decavèle, Marianne Friedman Polonsky, https://www.yonafriedman.org/.

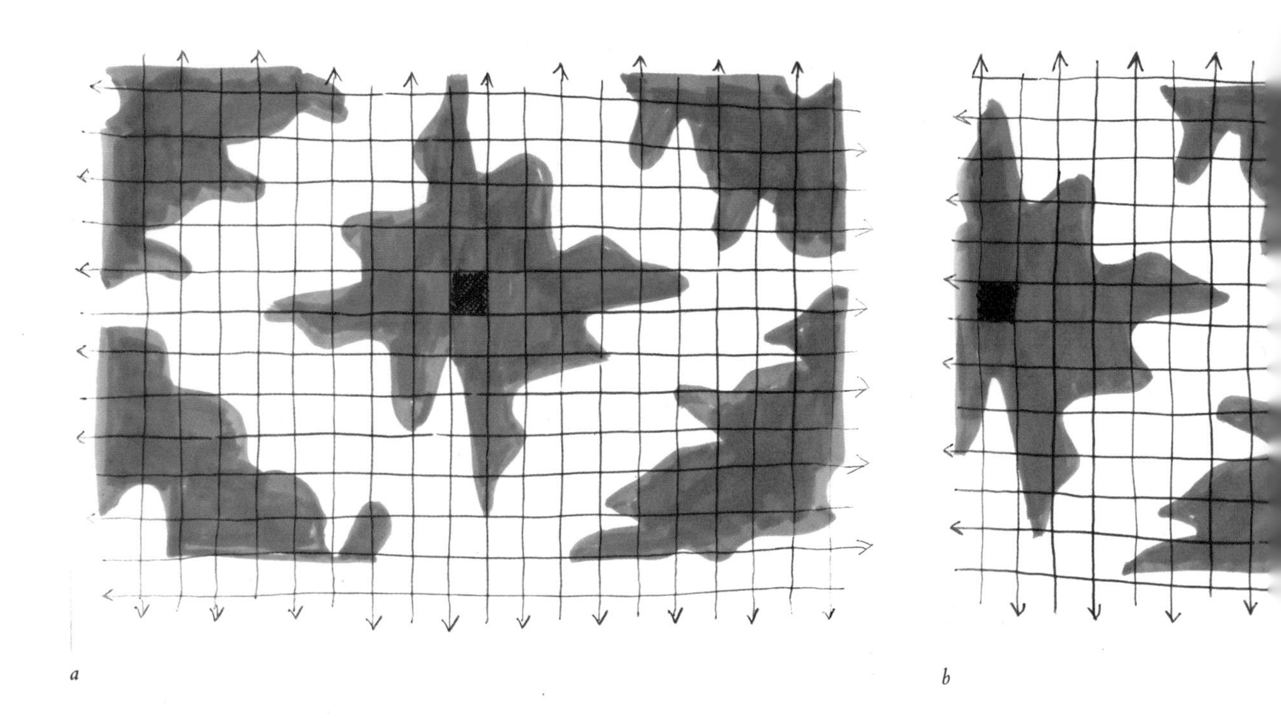

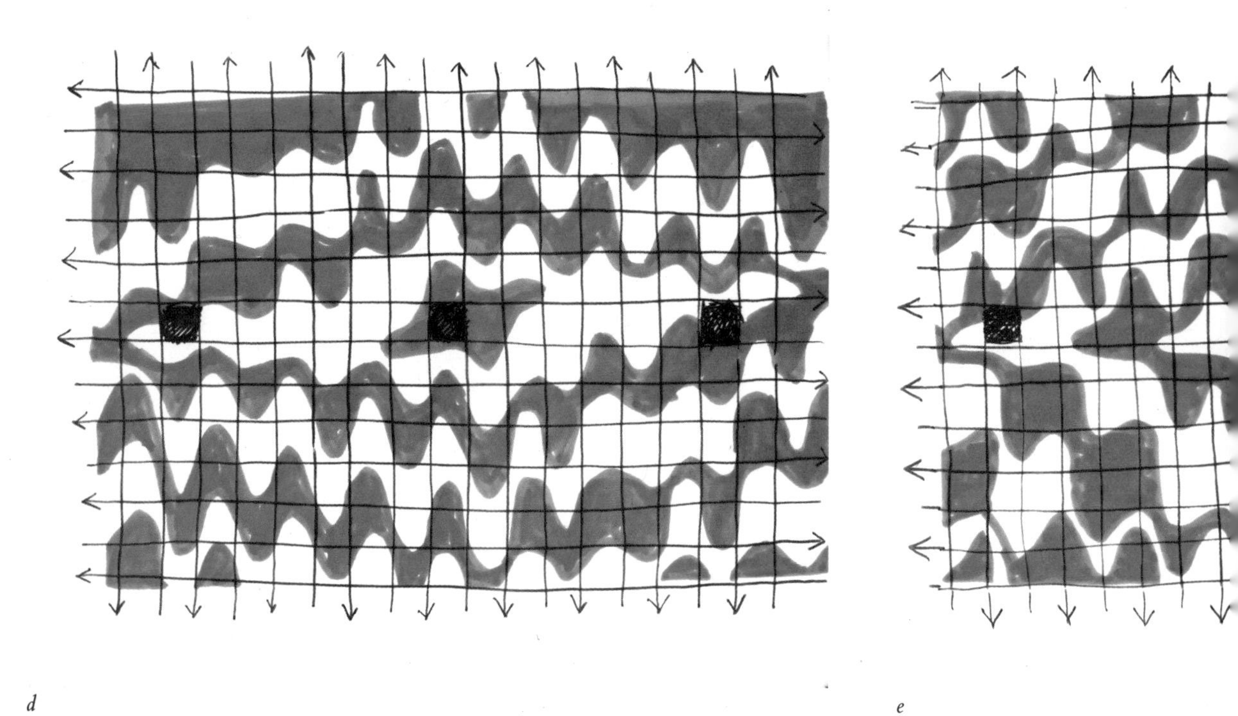

Figure 4.5
The future user: manual on architecture (n.d., circa late 1960). Sketches of "effort" distributions within an "urban mechanism." *Source*: Fond Denise et Yona Friedman, Folder 293, pages 253–258. Courtesy Fond Denise et Yona Friedman, Archives photographiques Jean-Baptiste Decavèle, Marianne Friedman Polonsky, https://www.yonafriedman.org/. All rights reserved.

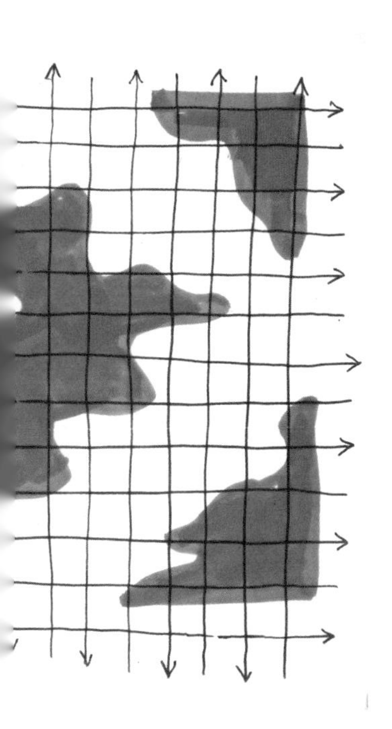

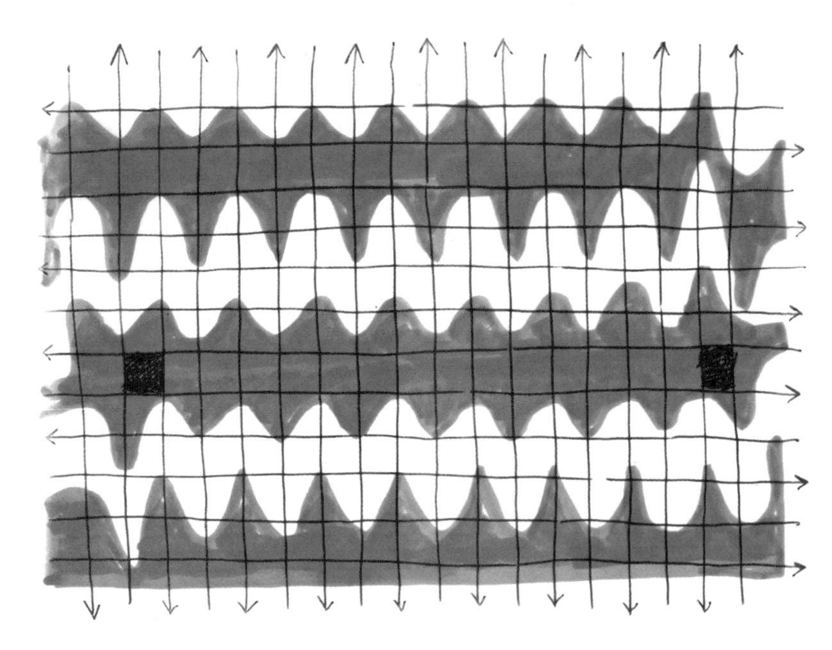

c

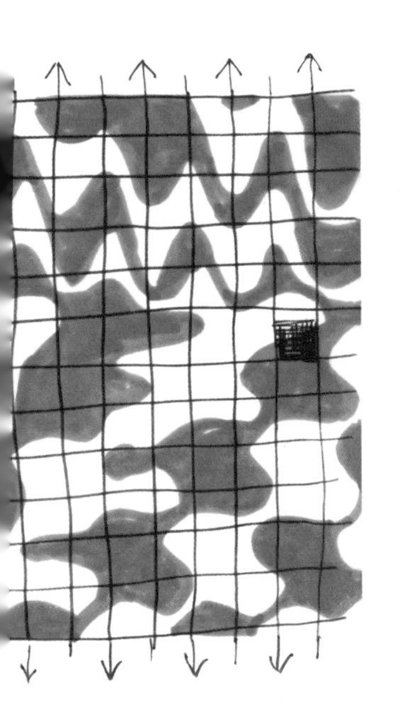

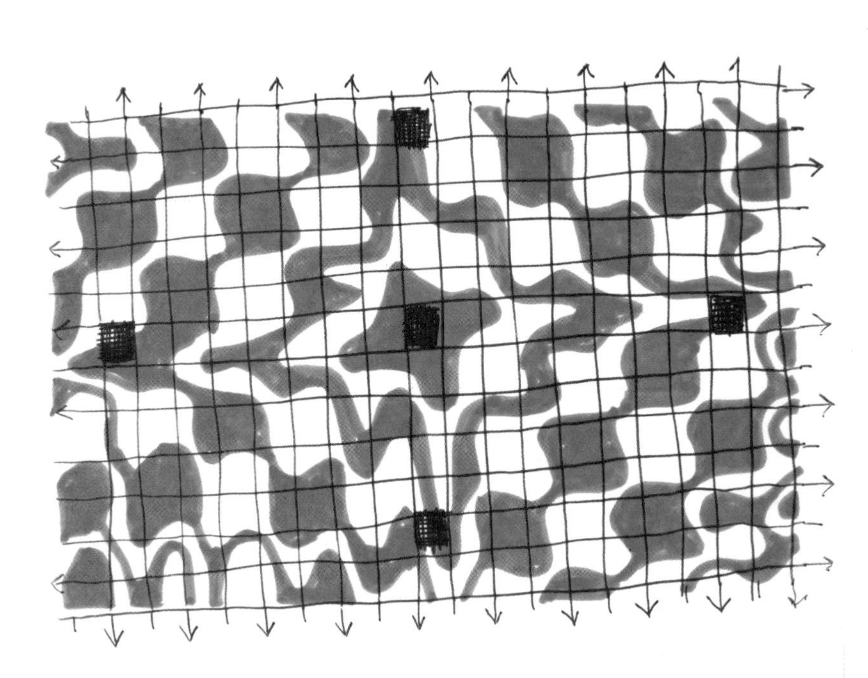

f

The machine also performed a "second loop," in which it calculated the urban-scale consequences of positioning the selected configuration in a specific location of the Ville Spatiale.[41]

The FLATWRITER machine was a remarkable amalgamation of computer-aided design techniques from operations research that Friedman had been exploring since the early 1960s and concerns that were at the forefront of urban planning realities in France. The mid-1960s found French planning organizations responding to growing discontent with the results of state-sponsored large-scale postwar housing initiatives, known as the *grands ensembles*, but also to emerging criticisms about conceptual and methodological flaws related to standardization and statistical reduction of individual and social life.[42] A new initiative to plan new towns, or *villes nouvelles*, together with an emphasis on architectural (as opposed to solely planning) research in the mid-1960s fostered proliferating discussions and experimental projects on dwelling types that would combine industry-mandated construction standardization with adaptivity to individual preferences and social changes.[43]

A key typology that emerged from these experiments was the *habitat évolutif*, a type of dwelling where inhabitants could choose from a *combinatoire*—a menu of combinations—of mass-produced building elements and standardized architectural components.[44] Despite its ambitions, *habitat évolutif* applications were limited. Actual implementations usually involved professional architects operating as "assistants," who informed users about the system of choices, sketched possible layouts according to their needs and aspirations, and finished the final apartment plan.[45] Friedman positioned the FLATWRITER as eliminating intervention from the "architect assistant," allowing its users to both self-operate the system of configurational choices and evaluate the consequences of a given configuration for their living habits and track their evolution.[46]

The use of a machine that could calculate and present, in what Friedman imagined as a thick book, the full range of choices along with a personalized metric enabling individual decision-making and collective deliberations seemed like a revolutionary response to the impasse of *habitat évolutif* projects. Or so claimed Michel Ragon, leading voice of radical architectural experiments in France, in his 1974 article "Yona Friedman: De l'habitat évolutif à l'autoplanification."[47] Ragon positioned the work of the dissolved GEAM, and in particular its founder Friedman, as the basis of the proliferating *habitat évolutif* experiments. He lamented that Friedman's ideas had been diffused to such an extent that they appeared "normal," despite the fact that they "ha[d] only been partially applied and always severely denaturing the initial point of view."[48] Although Friedman's work found limited application in the

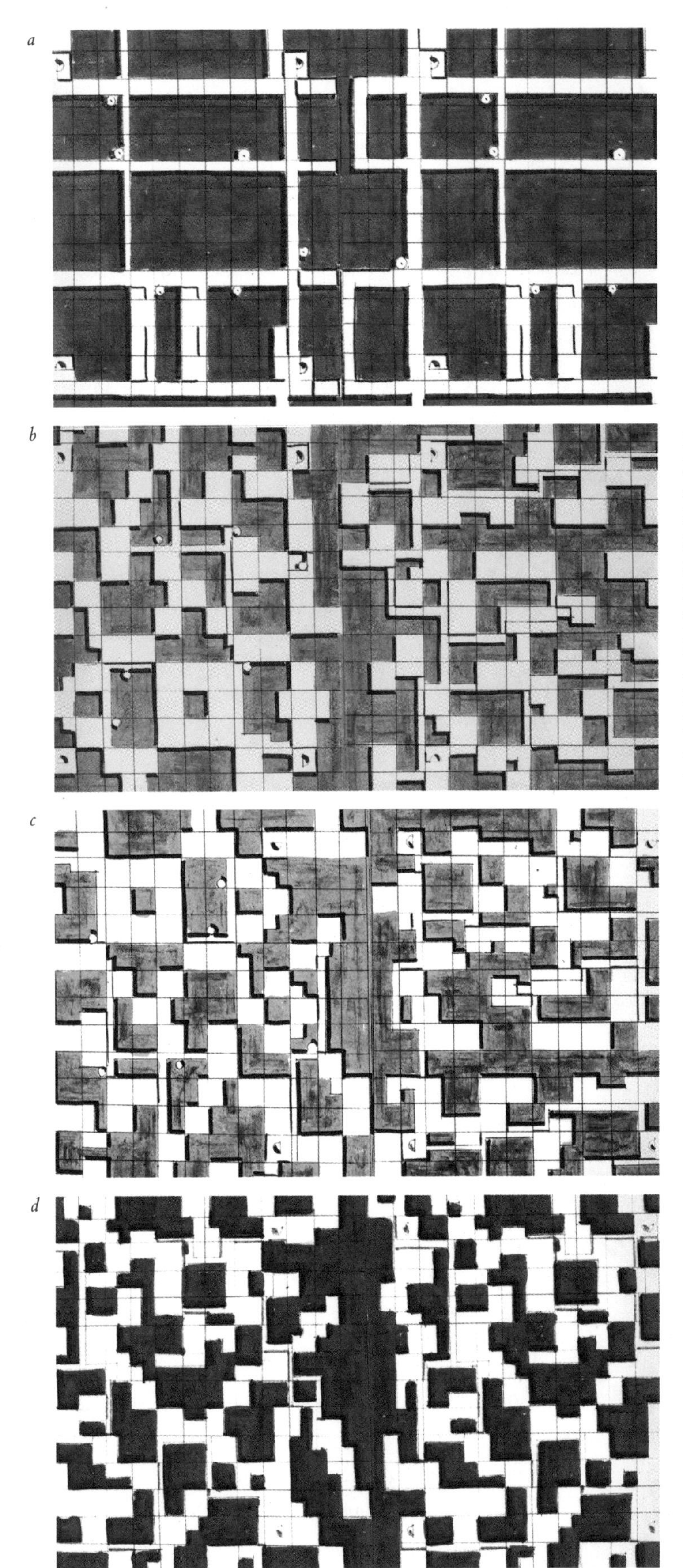

Figure 4.6
Massing plans of spatial neighborhoods (n.d., 1960s binder). Bottom neighborhood (*a*), interior neighborhood (*b*) and (*c*), top neighborhood (*d*). *Source*: Fond Denise et Yona Friedman, Folder 294, page 16–19. Courtesy Fond Denise et Yona Friedman, Archives photographiques Jean-Baptiste Decavèle, Marianne Friedman Polonsky, https://www.yonafriedman.org/. All rights reserved.

pragmatic context of French state planning initiatives, his theory of scientific architecture and his speculative FLATWRITER machine were met with eager interest by audiences working on design methods and computer-aided design in the United Kingdom and the United States.

Friedman presented his arguments for removing the professional "middleman" from the design process in a 1971 conference on Design Participation at the University of Manchester organized by the Design Research Society—the leading organization on design methods in the UK, founded in 1966 by members of the organizing committee of the 1962 Conference on Design Methods. The Design Participation conference was a response to ethical and philosophical self-scrutiny and demands for theoretical reorientation of British design methods after the 1967 conference Design Methods in Architecture organized by Geoffrey Broadbent and Tony Ward at Portsmouth.[49] The conference echoed boisterous criticism against scientistic tendencies that characterized design methods research, aiming to establish a distance from, as one *Architects' Journal* review described it, the "strictures" of "grotesquely oversimplifying problem solving structures."[50]

Motivated by such debates, Nigel Cross, active member of the British Architectural Students Association, had proposed to host a conference on participatory design at the University of Manchester, where he was a doctoral student at the Design Research Laboratory—a research group set up by Design Research Society founding member John Christopher Jones. Conference participants were designers, architects, and planners, convening with internationally acclaimed architectural critics (Reyner Banham being a notable example), artists, information technologists, and engineers.[51] Recurrent themes in the conference were a suspicion of technocracy and professionalism, questions of risk, expertise, decision-making legitimacy, and explorations of social and alternative technologies.[52] Friedman presented the FLATWRITER as the application of "Information Processes for Participatory Design."[53] These processes' ethical foundation was Friedman's motto that in a moral system, decisions ought to be made by the person who will be affected by their being wrong, by the person who takes the risk.

Also among the conference participants was Nicholas Negroponte, an alumnus of the MIT Department of Architecture who was hired as a professor shortly after graduating and founded the Architecture Machine in 1967. The facility benefited from MIT's timesharing computing systems developed jointly by MIT, General Electric, and Bell Labs in the mid-1960s, generous offers of minicomputers and peripherals from industrial donors, and custom sensors and actuators made in house by student researchers of varying academic levels and disciplines who passed through the Architecture Machine. With these facilities, and ample funding from sources such as the National Science

Foundation Division of Computer Research, the US Department of Defense's Advanced Research Projects Agency, and the Office for Naval Research, the Architecture Machine performed experiments to "assist, augment, and replicate design activities" by propelling "both the state-of-the-art of understanding design and by developing better hardware and software."[54]

Negroponte's paper at the Design Participation conference was titled "Aspects of Living in an Architecture Machine." It was a comparative review of varying attitudes toward computationally enhanced environments, specifically in relation to how they approach "learning" and "responding" to the inhabitant.[55] The review outlined some of the principles and motivations of a project on "intelligent environments" that the Architecture Machine had launched in 1969 with support from the Graham Foundation for Advanced Study in the Fine Arts and Ford Foundation funding through MIT. The intelligent environments research continued, and aimed to correct, earlier Architecture Machine experiments with behavioral modeling and environmental adaptation, exemplified by the infamous SEEK. Shown at the landmark exhibition "Software Information Technology: Its New Meaning for Art," curated by Jack Burnham at the Jewish Museum in New York in 1971, SEEK was a Plexiglas-encased blocks world of 480 metal-finished two-inch cubes, rearranged by an overhead pressure-sensing and electromagnetic robotic arm in response to the behavior of a colony of highly active Mongolian gerbils.[56] The Interdata Model 3 appended to the Plexiglas display evaluated discrepancies between its model of the block's configuration and gerbil-induced displacements. Depending on the displacement's magnitude, it ruled block dislocations either as accidents or "gerbil desired moves" and used these moves to devise a probabilistic model for where to place the next block.[57] Negroponte used SEEK to argue for the irreducibility of behavior to static models and the necessity of a mutual "learning" and adaptation among inhabitants and their environment.

The project on "intelligent environments" advanced ideas of conversation and evolutionary learning that were also present in the group's interactive computer-aided design system URBAN5 developed in 1967. The system staged a conversation between an architect and a machine that took the role of "an urban design clerk," evaluating various configurations proposed by the system user—an architect—against designer-specified criteria and hard-wired constraints.[58] URBAN5 allowed one to specify up to 100 general criteria in natural language, and it analyzed the sentences with a 100-word dictionary specific to urban design.[59] In the first conference of the Design Methods Group that was held at MIT, a major event with over 200 participants, Negroponte presented such expandability as a first step toward personalization.[60] "In just these examples of word-building," Negroponte wrote, "the designer is beginning to construct his own machine partner out of the aboriginal framework of URBAN5."[61] "This transformation," he continued, "occurs in the machine: the user is allowed to penetrate the surface of URBAN5, getting deeper and deeper into its assumptions and definitions"[62] The project on "intelligent environments" expanded these ideas of mutual adaptations between humans and machines through collaborations with cyberneticians Gordon Pask and Warren Brodey, from whose writings Negroponte readily appropriated the terms "soft architectures"—the title of the classic 1975 retrospective of the Architecture Machine's work—and "intelligent environments."[63]

It is unclear whether Friedman's and Negroponte's presentations at the Design Participation conference signal more than affine concerns and intersecting research networks. Could the event have rekindled a relationship that, as Negroponte wrote in the foreword to the English translation of *Pour une architecture scientifique*, began on a car ride they shared from Boston's Logan Airport while he was still a graduate student at MIT?[64] Friedman had certainly made an impression with his "soft spoken but persuasive argument," as Negroponte would put it, for giving users direct and unmediated control over the built environment because they were the risk bearers of "bad design."[65] Negroponte framed this kind of immediate control over one's built environment as the most radical form of participatory design and contrasted it with other more timid attempts at deprofessionalization presented at the Design Participation conference.[66] This affinity materialized in a research collaboration from 1973 to 1975, in which Friedman visited the Architecture Machine several times to consult on an incarnation of his theory in the computer system baptized YONA.

Choice

YONA was the computer implementation component of a larger experiment in computer-aided design titled Architecture-by-Yourself (often shortened as ABY), performed with the support of a research grant from the National Science Foundation Division of Computer Research as part of the project dubbed Machine Recognition and Inference Making in Computer Aids to Design.[67] The YONA system's developer Guy Weinzapfel presented the system with a mix of arguments on professional ethics ("unpracticed designers . . . unlike architects, bear a risk!") and fruitfulness for computer research ("the demands of the unpracticed will . . . accelerate development of graphical input techniques, display capabilities and design strategy systems").[68]

Weinzapfel suggested that because of their "demands to visualization" and "strategic assistance," computer aids for nonprofessional designers provided a "relentless setting" that could motivate further developments in computer graphics and computer-aided design.[69]

In a research-in-progress report that appeared in the July 1975 issue of the journal *Computer-Aided Design*, Negroponte was unapologetic about cashing in on do-it-yourself and participatory design philosophies to push computer research opportunities. He presented the Architecture-by-Yourself experiment, of which YONA was part, as "an application of principles in computer-aided design and of techniques in computer science which taxed each to their utmost," and continued to reassure readers that "while the philosophy of architecture-by-yourself may be distasteful to some, the techniques and systems have very real application to more pragmatic and professional views of architecture."[70]

Architecture-by-Yourself reused parts of other systems developed at the MIT School of Architecture and Planning.[71] There, at the start of the 1970s, computer research gathered around three research initiatives. The first, led by Timothy Johnson, was IMAGE, an automatic space-planning system that arranged rectangular volumes in three-dimensional space in order to reduce violations of designer-specified criteria and constraints.[72] The second, led by William Porter, was DISCOURSE, a programming language specific to problems of architecture and planning.[73] The third was the Architecture Machine. Weinzapfel had pursued research related to IMAGE as a graduate student in the MIT Department of Architecture and integrated parts of IMAGE in the Architecture-by-Yourself project.[74] His work with IMAGE led him "quite naturally," as he put it, to the computer implementation of Friedman's "graph theory approach" in the YONA system.[75] IMAGE cast design as a process of evaluating elementary architectural *configurations* based on how they performed in relation to objectives. A computer-aided design system for the "unpracticed" such as YONA, Weinzapfel noted, required a representation that could describe both the design and the designer's objectives. By representing "the linkages of the house—the connections [doorways] between its spaces," Friedman's graph-based method "provide[d] both the backbone of the design and the largest payoff in terms of describing the designer's objectives."[76]

In YONA, Weinzapfel proposed the graph both as a malleable representation of architectural arrangement and as a calculative device. Instead of hiding it as background computation, YONA visualized the graph on a touch-sensitive computer screen and allowed the users to manipulate it using their fingers (instead of the light pen, which was the customary input device at the time).[77] In his paper on YONA, Weinzapfel referred to an observational study of the "design 'actions'" of one couple of "unpracticed" designers over the course of eight weeks. According to Weinzapfel, these observations confirmed "visualization" and "strategic assistance" as crucial requirements. Weinzapfel eulogized the graph's many benefits in meeting these requirements: its versatility in capturing under one singular mathematical representation different stages of the design process, its simplicity, and its informativeness in revealing otherwise unseen conflicts of connectivity, access, or enclosure[78] Most salient among these many traits, and specific to the way that Friedman construed the graph, was its intuitiveness, Weinzapfel wrote: "Friedman's approach decomposes the design problem, not by abstract concepts of form and function, or by service spaces and areas served, but by the more tangible factors of room placement, connectivity, size, shape and 3D form."[79]

Architecture-by-Yourself and the YONA system were based on technical and discursive threads that the Architecture Machine had been pursuing since the start of the 1970s. In that context, the emerging architectural agenda of "participation" served as a premise for furthering research on personalization and human-machine evolutionary learning that was already present in URBAN5 and the intelligent environments work. For instance, after Negroponte presented URBAN5 in the 1968 Design Methods Group conference at MIT, he received an enthusiastic response from computer artist and Bell Labs researcher Michael Noll, who was a respondent for the Computer-Aided Design section of the Design Methods Group conference. Noll was captivated by Negroponte's suggestive discussion of mutual learning and evolutionary personalization between the human designer and a "self-teaching system" such as URBAN5.[80] This seemed to suggest the possibility of giving those "affected by city planning and architectural environments . . . a stronger role and perhaps even the opportunity to do some of the designing themselves."[81] Negroponte swiftly embarked on the idea and embraced it as one of Architecture Machine's inaugural pursuits.

Transformative events at MIT such as the establishment of the Black Student Union, student activism, urban poverty and community housing studies by faculty members and visitors such as Robert Goodman, John Turner, and Giancarlo De Carlo, and the release of Stanford Anderson's *Planning for Diversity and Choice* paint part of the context in which Negroponte's attitude toward computer aids shifted toward ideas of participation.[82] A concern with the future, along with debates on ethics, power, and professional responsibility on the one hand and modeling, prediction, and anticipation on the other, increasingly preoccupied designers in the late 1960s. Evidence of this were several committees and research projects on the topic, mainly in the social sciences, that had an impact on architectural and environmental design research.[83]

Negroponte described the Architecture Machine's turn toward participation as a shift from "emulating [architects] in computers" to produce a "super architect, a surrogate architect" to "doing away with architects [in housing]" and developing a "computer aided instruction system and mediator for the goals of the group and the desires of the individual."[84] The first outline of the computer implementation of such a process appeared in a proposal that Negroponte submitted to the National Science Foundation in 1971, together with Leon Groisser, with the title "Computer Aids to Participatory Architecture." The research proposal received a $60,500 two-year award (1972–1974) from the National Science Foundation's Institutional Support for Science and Computing Activities program and formed preparatory ground for the Architecture-by-Yourself project. Sharing common ground with URBAN5, especially as it pertained to human-machine communication, the 1971 proposal described a "straightforward, free-wheeling, and congenial conversation" between the "proverbial man-in-the-street, who is both a novice in architecture and a novice with machines," and "a machine which has some 'knowledge' of architecture . . . not only appear[ing] to be a 'competent' architect, but . . . a sympathetic conversant, a good model builder, graphically dextrous, and friendly."[85] Such a machine would provide "the necessary design tools" to enable "each man being his own architect or . . . each man at least participating in the design of his immediate built environment."[86]

Negroponte and Groisser conceptualized the "dweller's plot of land" as a three-dimensional abstract spatial unit.[87] Depicted in the proposal with a drawing of wireframe parallelepipeds superimposed with graphs connecting their centers of gravity, this abstraction was reminiscent both of URBAN5's "frictionless vacuum" of cubes and of Friedman's renditions of the spatial city's infill.[88] Friedman's ideas were also present in the phasing of the research proposal, which Negroponte and Groisser presented as taking place in two rounds, one "concerned primarily with the design of (the dweller's share of the chunk of space) and one addressing "the problem of control and mediation of the individual's needs and desires with respect to his neighborhood."[89]

The system would build on IMAGE, from which it would derive methods for constraint and criteria resolution, and on DISCOURSE, which it would use for its extensive data structure work. The new system, however, would aspire to personalization and congeniality—elements that Negroponte and Groisser saw as lacking in other computer-aided design systems.[90] Interactional fluidity was in keeping with the "thrusts of the Architecture Machine Group": "interfacing with a user and dealing directly with the real world and the procedures for handling problems of context and of missing information."[91]

Unlike systems like IMAGE or other automatic space planning systems that took as input a well-defined set of constraints and criteria, the system Negroponte and Groisser proposed aimed at inferring "stated or unstated criteria" from its user's design acts.[92] This marked a conceptual shift from "specification" to "recognition."[93] Such "recognition" required an intelligent machine that could go beyond the information provided and reason about its meaning based on the information's context—in this case, an evolving model of the system's user. The system elicited information about its user's design criteria by "*observing the relationships* [emphasis mine] that the user establishes graphically and linguistically."[94]

In order to pick up these "graphical relationships," the system would use sketch recognition—a topic that was occupying much of the Architecture Machine's research activity at that time.[95] For example, in 1971 the Architecture Machine developed a sketch recognition system named HUNCH. HUNCH included a syntax of elements for representing sketches hand-drawn on a Sylvania data tablet.[96] Negroponte contrasted HUNCH with the "rubber-band pointing-and-tracking vernacular" promoted by systems such as Ivan Sutherland's landmark computer graphics system SKETCHPAD. SKETCHPAD, Negroponte lamented, "polluted the notion of 'sketching', in any sense of the word," by idealizing elements drawn by its users as pristine geometric forms and structures.[97] HUNCH, on the other hand, saved gestural information about the sketches—"the wobbliness of lines, the collections of over tracings, and the darkness of inscriptions."[98] It took in, as Negroponte described, "every nick and bump, storing a voluminous history of your tracings on both magnetic tape and storage tube."[99] With that data, the system made inferences about the drawing's geometric meaning, the designer's intentions, and even preferences in architectural style.

In the "Computer Aids to Participatory Architecture" funding proposal, Negroponte and Groisser suggested the use of SQUINT—an offspring of HUNCH—that could recognize boundaries and interpenetrations of shapes. In doing so, SQUINT could recognize "positional" (direction and orientation) and "proximity" (adjacencies, connections, overlaps) relationships implicit in a hand-drawn sketch of a house plan made by the system's architectural novice user. Interpreted as unstated criteria, such relationships were topological and nonmetric—they were graphs.[100] Graphs provided "an initial, though crude, overview of the user's criteria," while ensuring congruence with the generated designs.[101] After extracting an "architectural program" from its user's sketches and representing it in the form of a graph, the system would then use the graph to generate alternative floor plan configurations. The system did not output an optimum or suboptimum layout nor a full list of alternative options. Instead, the system and its user engaged in cyclical interactions where the machine "generat[ed] alternatives and encourage[ed] the user to alter arrangements" while "storing and retrieving 'apparently satisfactory states'."[102] The goal was for the user and the machine to converge together toward more appropriate layouts.[103]

Through these interactions the system was not only modeling its users' preferences in architectural configuration but was modeling the users themselves. Or as Negroponte and Groisser put it, the system was "build[ing] a model of the user's new or modified habitat" while "simultaneously building a model of the user and a model of the user's model of it."[104] The intended result of this process was to produce physical form and express it graphically but also to produce self-awareness ("discover, understand and express his own needs and desires," "to understand more fully his own patterns of living and how they affect and are affected by the physical environment in which he lives [underlined in the original]").[105] Such understanding did not grow out of the conscious self-tracking of daily habits as Friedman imagined in the FLATWRITER, but in conversational immersion with a machine that inferred intentions from its user's sketches and outputted corresponding architectural proposals. The Architecture Machine, put differently, displaced the graph from an infrastructure of possibility to an infrastructure for human-machine communication, a medium that permitted mappings between personal utterances and the machine's objective calculations.

Possibility

As the Architecture Machine was building computer aids for do-it-yourself architecture with Friedman's graph-based methods, Lionel March and Philip Steadman, together with a growing network of mathematically inclined researchers, were beginning to develop the technical and institutional frameworks for establishing a "science of possibility" in architecture.[106] Although intellectually akin to their work, systems such as the FLATWRITER left them unimpressed. In *Architectural Morphology*, for instance, a 1983 book documenting principles and mathematical techniques that had emerged from these forays in architectural possibility, Steadman decried the FLATWRITER's rote combinatorics. "The flatwriter," he wrote, "is a kind of architectural counterpart of the proverbial monkeys typing Shakespeare, or the Lullian combinatorial machines for literary composition which Gulliver found at the Academy of Lagado."[107] Steadman also took issue with Friedman's "eccentric and erratic series of graph-theoretic ideas" and his "unreliable arithmetic."[108] The FLATWRITER miscalculated the possible combinations by a factor of six. The frequency of "Mr.

X's"—Friedman's "future user's"—movements was also wrong. The number of times Mr. X was entering a room was different from the times leaving it.[109]

The mathematical armature of Steadman's "general *science of possible forms* [emphasis in the original]," as he defined "morphology," also largely consisted of graphs as infrastructures on which concrete architectural layout configurations could be calculated.[110] Steadman had engaged these ideas while working on automating minimum-standard floor plans. The work refrained from bold social visions and ruminations on the politics of architectural decision-making that permeated Friedman's work and were coopted by the Architecture Machine. However, navigating vast ranges of options within fixed constraints hinted at new prospects for individual expression within governmental imperatives of standardization in design and construction.

Steadman discussed these prospects in his 1973 paper "Graph Theoretic Representation of Architectural Arrangement," originally published in *Architectural Research and Teaching* and reprinted in March's 1976 edited collection *The Architecture of Form*.[111] Developing a mathematical method for counting all possible floor plans could motivate a wide range of experiments, from precisely correlating governmentally set housing standards with a variety of allowable plans to exploring relationships between prefabrication systems and spatial arrangement possibilities. Lionel March juxtaposed the elaboration of possibility to the "'tight fit' functionalism" that imagined architectural geometry *determined* by functional constraints. A clear understanding of architectural possibility would ensure, he claimed, that "a 'loose fit' approach is not sloppy but, on the contrary, as well tailored as a good off-the-peg suit."[112]

Steadman referred to the exhaustive counting of architectural layouts as the "enumeration" approach to representing architectural arrangements through graphs. Unlike the "heuristic" approach that used graphs to develop an architectural layout that met a set of adjacency or proximity requirements—as we saw, for example, Levin counterpropose to Whitehead and Eldars's automatic layout generation computer program—the "enumeration" approach computed *all possible layouts* for a set of constraints. Steadman mused about enumeration being "perhaps intellectually more interesting and elegant" by virtue of being exhaustive.[113] "The [computer] program," he wrote, "need only be run successfully once for any particular problem, and *then the results are known* [emphasis in the original]."[114] Enumeration could identify "*all* feasible solutions [emphasis in the original]," "once and for all."[115] This definitive record of possibilities could be circulated as abstract, schematic layouts, possibly in book form. Architects would elaborate these schemes to fully fledged floor plans.

In 1976 Steadman collaborated with William (Bill) Mitchell and Robin Liggett from the School of Architecture at the University of California, Los Angeles (UCLA), on a computer algorithm for counting and optimizing floor plans. Mitchell, who had joined UCLA after completing a master of environmental design at Yale University, was also affiliated with the University of Cambridge from where he earned a master of arts degree in 1977. Liggett was a mathematician who was pursuing doctoral studies in operations research at the UCLA School of Management and teaching related courses in the School of Architecture. The outcome of the collaboration was a package of FORTRAN computer programs tested on UCLA's IBM 360/91 computer and on the IBM 370/155 computer at the Yale University Computer Center. These systems' speed established clear limits for the algorithm's capabilities. The IBM 370/155 at the Yale University Computer Center needed about one hour to exhaustively enumerate all floor plan possibilities for eight rooms. Nine rooms, generating about 250,000 distinct schematic plans, were on the border of being unfeasible, and ten were "out of the question."[116]

The foundation of the program was a "special technique" for representing floor plans that the authors referred to as a "dimensionless representation."[117] This involved surrounding every plan with a bounding rectangle, passing horizontal and vertical lines from all vertices of the plan so that each vertex lies on an intersection of these lines, and modifying the areas defined by these lines so that they are all squares. This "minimum rectangular grating," as the dimensionless representation was called, preserved the floor plan's topology and could be used to generate infinite variations of topologically equivalent plans by modifying the dimensions of the shapes. This representation was ideal for computer input. A grating of 5 × 3 cells, for instance, would correspond to a matrix of 5 columns and 3 rows. Each position in the matrix could be assigned a number representing the function of the cell (room) in the grating. An x vector of 5 numbers (the number of columns) and a y vector of 3 numbers (the number of rows) would then store the dimensions of each cell.[118] The matrix and the two vectors could provide a full description of a rectilinear architectural floor plan. The matrix represented the plan's topology. The vectors represented its dimensions. This allowed, as the authors celebrated in the article, separating algorithms that treated the plan's topology from those treating its dimensions.[119]

The 1976 computer program used rectangular dissections to count all topologically distinct floor plans for a set number of rooms. Assigning room types to these schematic floor plans' cells required representing them as graphs. The task was then to establish correspondences between the "adjacencies requirements matrix"—indicating

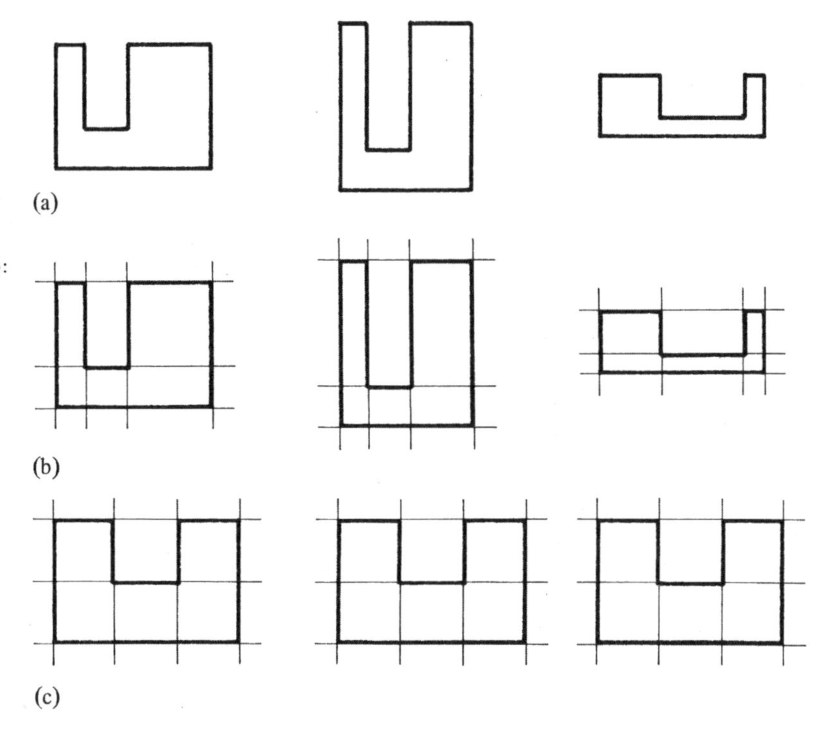

Figure 4.7
Unifying the representation of dissimilar forms through the same "dimensionless shape."
Source: William Mitchell, Philip Steadman, and Robin Liggett, "Synthesis and Optimization of Small Rectangular Floor Plans," *Environment and Planning B* 3 (1976): 38. © 1976 SAGE.

relationships between rooms and orientation—which could also be expressed as a graph.[120] The otherwise arduous and uneconomical assignment became efficient through various tests for ruling out non-planar adjacency graphs, graphs that could not be realized as a rectangular dissection, or dissection graphs whose vertex orders (numbers of connections) exceeded those of the adjacency graph.[121]

UCLA's IBM 360/91 could assign rooms to dissections of six or fewer cells within seconds. Computing time increased rapidly when adding more rooms, with seven rooms requiring under a minute and eight rooms needing several minutes to compute. After the adjacency assignments, a set of algorithms gave dimensions to the rooms based on dimension, area, and proportion constraints. The x and y vectors contained the data for this step. The paper described methods for superimposing metric considerations on the realm of relations represented by the "non-dimensional" floor plans. Recall Steadman's electrical circuit analogy and his graph-based method for assigning dimensions to floor plans. That method was based on standard minimum room dimensions. The 1976 paper took on the more taxing task of optimizing dimensional assignments based on objectives such as minimizing or maximizing total floor area or perimeter-to-area ratio.[122]

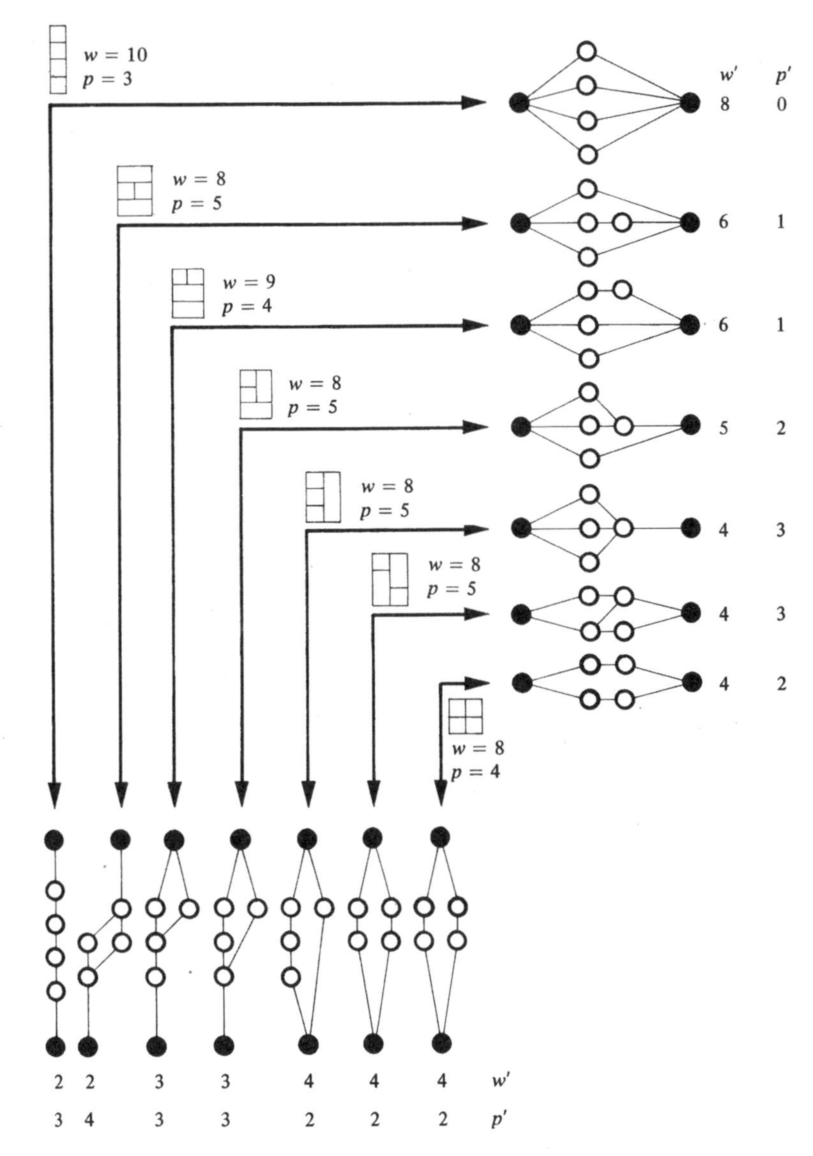

Figure 4.8
Seven dissections for four rooms and their corresponding "half-graphs" indicating adjacency relationships in the horizontal and perpendicular direction.
Source: William Mitchell, Philip Steadman, and Robin Liggett, "Synthesis and Optimization of Small Rectangular Floor Plans," *Environment and Planning B* 3 (1976): 50. © 1976 SAGE.

The elaboration of architectural possibility was intimately linked with representation. As we saw, March and Steadman discussed at length the promises of new mathematical ideas in expanding architectural representation in *The Geometry of Environment*. When the book was written in 1971, its intellectually tantalizing collection of new mathematical ideas did not yet amount to a distinct agenda for design and architecture. A few years later, this agenda began crystallizing under the keyword of "configuration." In a joint presentation they gave in 1978 at an International Conference on Descriptive Geometry

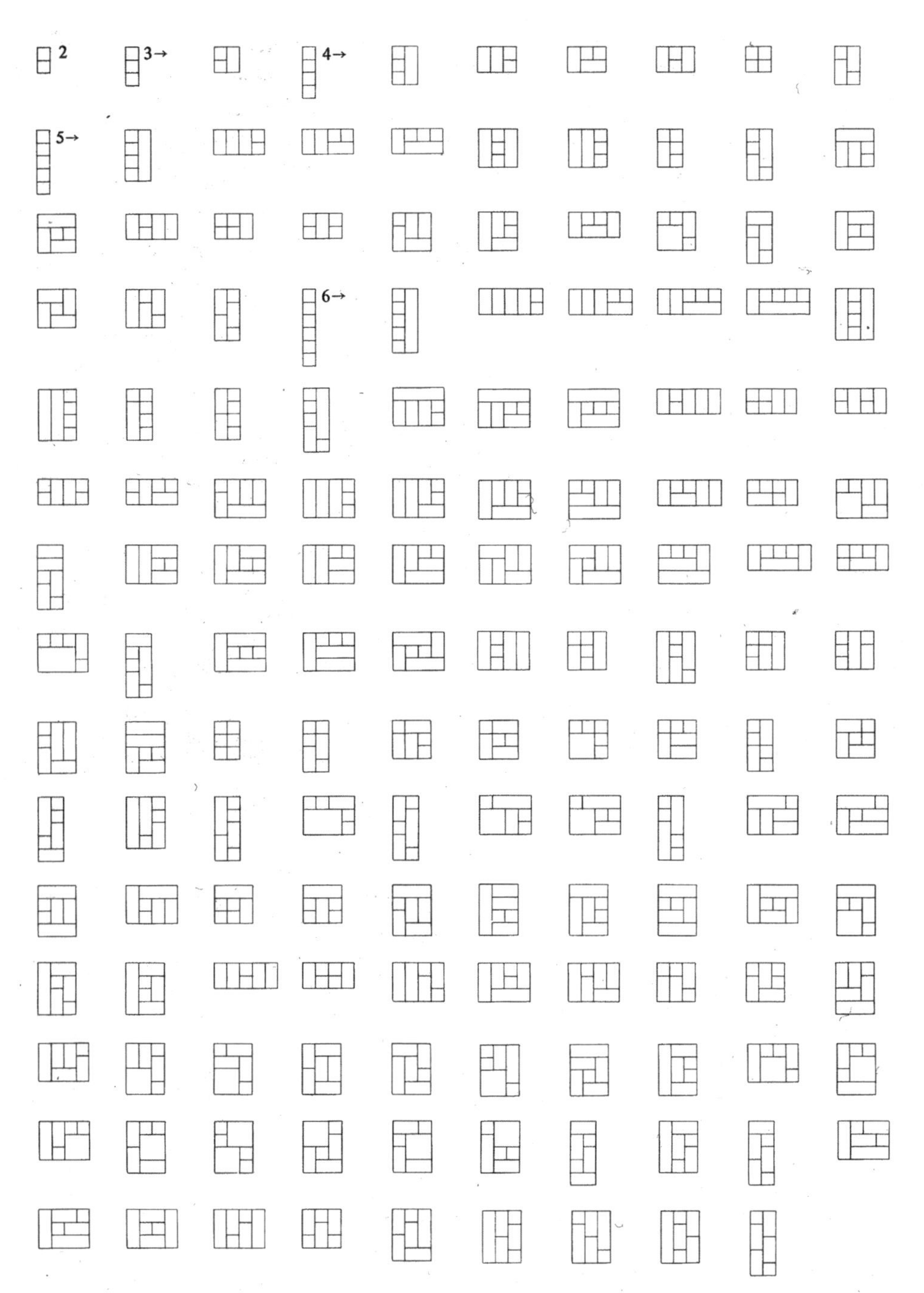

Figure 4.9
Plot of dissections for up to six rooms. *Source*: William Mitchell, Philip Steadman, and Robin Liggett, “Synthesis and Optimization of Small Rectangular Floor Plans,” *Environment and Planning B* 3 (1976): 70. © 1976 SAGE.

in Vancouver—the inaugural event of the International Society for Geometry and Graphics, still active today—March and Steadman declared descriptive geometry's "academic demise" as a mere tool for *picturing* objects, a direction they argued also characterized computer graphics research.[123] They vowed instead to reclaim geometry's "constructive" potential: its ability to be used in the elaboration of design possibilities. This would be achieved through what they called "configurational engineering."[124]

"Configurational engineering" included processes of generating, searching, selecting, and optimizing the form of objects in appropriate representation spaces, and then the transformation of representations to produce metric descriptions of the object. Transformations between representation spaces were, as we have seen, a fond topic in British new math debates about geometry and had received keen attention in *The Geometry of Environment*. In discussing the hierarchy of representations in their 1978 paper, March and Steadman observed that relaxing the invariance of collineation in projective geometry led to topology. The field of "pure structure," as they described it, was a felicitous realm for combinatorial generation and evaluation of design possibilities.[125]

Soon after the Vancouver conference, during his tenure as chair of design at the Open University, March established the Centre for Configurational Studies as the research division of the Faculty of Technology's "Design Discipline" (the equivalent of a department). The Open University was a distance-learning institution established in 1969, with a physical campus dedicated to research at Milton Keynes. The Design Discipline was at the forefront of research on systematic approaches to design, with John Christopher Jones becoming its first chair, and benefited from state-of-the-art computing facilities such as the VAX 11/780 system at the end of the 1970s.[126] The Centre for Configurational Studies' first triennial report presented a shared concern with "questions of form, pattern, and structure in design and planning" as the thread that tied together the Centre's varied projects and agendas that spanned from architecture to robotics and from the history of Soviet design and planning to biology.[127] The report earnestly described these loose connections as results of the Centre's *history*: the Centre was about the coming together of kindred researchers who had crossed paths over the years. Upon founding the Centre, March invited past students from the Land Use Built Form Studies Centre, such as Catherine Cooke and Philip Steadman, as well as research supervisees from a visiting appointment at the University of Waterloo Department of Systems Design from 1974 to 1976.

March's short stint at Waterloo was essential for the development of a focus on architectural possibility, one of the Centre for Configurational Studies' key foci. In the mid-1970s the University of Waterloo had already garnered an international reputation as a leader in science and technology. The university hosted state-of-the-art computing facilities, developed under the stewardship of mathematics professor and former IBM engineer Wes Graham.[128] Its Faculty of Mathematics, the first independent faculty on the subject in North America, had attracted world leaders in combinatorics and statistics including Bill Tutte—a member of Blanche.[129] The university's Department of Systems Design Engineering was also in the spotlight. Founded in 1968 by Department of Design faculty member George Soulis, together with Peter Roe and H. K. Kesavan, using funds collected from work they performed for the 1967 International and Universal Exposition (Expo '67) in Montreal, the department was a central hub for work on systematic design processes and methods.[130] This growing reputation solidified a position that the University of Waterloo had assumed through impactful events such as the 1966 International Conference on Design and Planning themed "Computers in Design and Communication."[131] Aside from resulting in the publication "Design and the Computer," a special issue of *Design Quarterly* coedited by Peter Seitz and Martin Krampen and known for its iconic cover of scattered letters emulating a punched card, the conference became known as the founding event of the Design Methods Group—the preeminent North American scholarly organization for design methods activity.[132]

The context of the University of Waterloo, host to many visiting researchers from the UK and North America, helped March forge a network essential for the Centre for Configurational Studies. For example, March collaborated on predictive land use models for the Toronto region with Michael Batty, then visiting assistant professor in the University of Waterloo Department of Civil Engineering. Batty would become a leading figure in urban modeling and editor of *Environment and Planning B: Planning and Design*—a foremost journal in architecture, design, and mathematics and the leading dissemination medium for the work of March's network of supervisees and affiliate researchers.

At Waterloo, March also found a prime site for unapologetically pursuing his mathematical exploration in architectural form and consolidating an agenda around the science of architectural possibility. Together with his research supervisee Ray Matela, a PhD graduate from Yale whom he later invited as one of the Centre for Configurational Studies' charter members, March developed work on "polyominoes"—planar arrangements of squares connected at an edge.[133] A helpful way to visualize this is Tetris blocks (incidentally grounded on polyominoes). The term "polyomino" was introduced by Solomon Golomb in the *American Mathematical Monthly*, and its equivalent, "animal" was coined by R. C. Read to describe issues of cell growth.[134]

In the architectural applications of polyominoes March and Matela saw the foundation for a systematic understanding of architectural form. Polyominoes, they argued, could represent architectural floor plans, with every room abstracted as a planar square face—a cell. The number of configurations of these planar square arrangements exploded to astronomic heights when the number of faces increased: for example, while three faces gave two distinct shapes, six gave 35. Arranging nine rooms gave 1,285 possibilities of different configurations, and 18 rooms exploded the collection to 192,622,052 distinct arrangements.[135] An architect working with a brief of 18 rooms, put differently, was choosing between some two hundred million possibilities. Counting and mapping out this startling variety was an enthralling undertaking.

March and Matela presented the motivations and technical underpinnings of a "science of form" in a 1974 article published in *Environment and Planning B* with the suggestive title "The Animals of

Figure 4.10
A part of the population of 369 8-ominoes. *Source*: Lionel March and Ray Matela, "The Animals of Architecture: Some Census Results on N-omino populations for N = 6, 7, 8," *Environment and Planning B* 1 (1974): 195. © 1974 SAGE.

Architecture." The title was directly taken from an eponymous 1972 paper written by University of Waterloo systems engineering design researchers Rammohan Ragade and Peter Roe, together with Robert Frew who, after training in the same department, had joined the faculty in the School of Architecture at Yale University.[136] Frew, Ragade, and Roe presented the paper at the third conference of the Environmental Design Research Association (EDRA 3)—an offshoot of the Design Methods Group that included a stronger presence of social and behavioral scientists. The conference, held at the UCLA School of Architecture and Urban Design in January 1972, was organized jointly with American Institute of Architects (AIA)-Architect-Researchers—an organization founded in 1964 with a focus on new design and building methods and techniques.[137]

Frew, Ragade, and Roe's paper focused on translating a set of distances set by a client into an optimal spatial layout. The layout was in the form of a polyomino "pattern," a two-dimensional array of square cells. Motivating the paper were aphorisms about architects' roles as educators and the consideration of a client's "inherent values" as what separates architects from salespeople.[138] "Values," in the context of the paper, were client preferences that could be expressed as numbers. The client, for instance, would decide the number of cells to form the overall pattern and how close or far a cell should be in relation to other cells. The authors used graphs to represent these distances and polyominoes of three cells, called "trominoes," to construct a gradually growing pattern that satisfied as much as possible the distances determined by clients.[139] This process, which the authors referred to as "educative" and "diagnostic," would allow for the study of preference and decision-shifting, which Frew was investigating at Waterloo.

March and Matela's paper was institutionally and intellectually contiguous but forbore proposing a design method. Although it didn't shy away from provocative positions about the architect's role and the future of the discipline, the paper, in typical March fashion, dwelled on education and research. In an also familiar way, the authors rejoiced in describing comprehensively and educatively fundamental concepts and properties of polyominoes. The goal was not to introduce a new actionable method, but to motivate new research. Among the paper's key points was the polyominoes' representational versatility that allowed one to move between algebraic, geometric, and topological descriptions, namely the matrix of numbers, the "map," and the graph.[140] The "map" was generated by taking the center of each polyomino (a point corresponding to each face) and adding a set of four planar points so that a square unit was created around the center. The graph, in turn, was defined by taking the center of each polyomino as a vertex of the graph and joining the vertices by edges so that no more

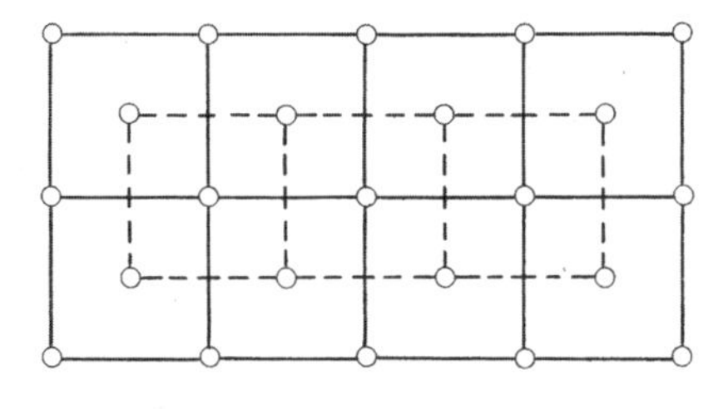

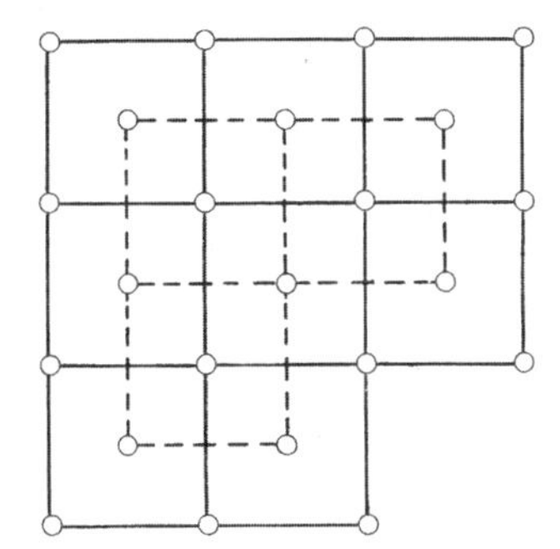

Figure 4.11
Two 8-ominoes and their graphs. *Source*: Lionel March and Ray Matela, "The Animals of Architecture: Some Census Results on N-omino populations for N = 6, 7, 8," *Environment and Planning B* 1 (1974): 201. © 1974 SAGE.

than four edges joined at a vertex. March and Matela reminded readers that the graph was immune to geometric considerations and could be drawn in any way deemed "convenient" or "expressive."[141] Using these representations, they defined a roster of related notions such as the "cover" of the polyomino, density, proportions, and distances—all corresponding to concepts of architectural significance that, as we saw from the EDRA 3 conference, other researchers had already taken on.

Following this overview, teeming with mathematical formalisms and compelling visuals, March and Matela discussed the position of systematic knowledge on form for architectural education and knowledge.[142] They juxtaposed the mathematical investigation of form to the patent emphasis of contemporaneous architectural research on sociological and behavioral affairs. Form, they argued, remained delegated to the "artist" and not the "scientist," resulting in an unfortunate situation in which "the science of form," as they outlined the predicament, "is not developed and supported because the science of form is not developed enough to be supported."[143] March and Matela argued that a robust vocabulary of form was a prerequisite for any functional considerations and critiqued attempts to "let form follow function in some supposedly self-generative way" (think, for example, of work on automatic floor plan generation based on pedestrian traffic that, as we know well by now, had dominated much of computer-aided design research since the early 1960s).[144] March and Matela pointed out that such automatic spatial synthesis work *used* polyominoes without knowing about them or their mathematical properties (remember the square grid cells arranged in Whitehead and Eldars and their material predecessors in systematic layout planning manual methods).

The language and properties of polyominoes had enabled architectural researchers to expand similar methods as tools for exploration as opposed to optimization. For instance, at EDRA 3, conference chair Bill Mitchell together with Robert Dillon, computer services coordinator of the Los Angeles firm Welton Becket and Associates, outlined a computer program called DOMINO.[145] DOMINO was a batch program with line-printer output that was also being developed as an interactive program on a PDP-10 at UCLA. Welton Becket and Associates used the program on commercial and hospital building design projects to generate preliminary, "rough" ideas on floor plan layouts. DOMINO's premise was that floor plan layout diagrams, when mapped into two-dimensional arrays of integers, were in fact polyominoes.[146] Mapping a drawing onto an array of integers was a prime method for its computational encoding. A grid was overlaid on the drawing and its lines were adjusted to fit the closest lines on the grid. This was translated to a two-dimensional array of numbers where 0 was outside the drawing outline and 1, 2, 3, 4, 5 . . . represented distinct areas inside the drawing outline. DOMINO's aim was to offer, in a fast and economic way, a wide range of alternatives that architects could then architecturally explore, alongside useful evaluation metrics for these alternatives. The program worked additively, "growing" polyomino patterns by starting with a single square and adding one at a time so that every square shares at least one side with an already existing square. The addition considered adjacencies, areas, and proportions of the departments.

Such practical computer applications were beside the point for March and Matela, who saw in polyominoes the possibility of achieving a deep and comprehensive mathematical understanding of architectural arrangement. Characterization of these arrangements' geometric and topological properties would provide a bedrock for functional reasoning and would distinguish architects from other researchers approaching the built environment from an anthropological or sociological standpoint. Ultimately, it would render geometry—in the sense put forward in March and Steadman's earlier articulation of "configurational engineering"—a priority in architectural decision-making, one pithily synopsized in the adage "look before you leap."[147]

In 1978 March was awarded a £17,394 grant from the UK Science Research Council for a project on "Systems Representation of Architectural Room Layout Planning," which he developed with Joseph Rooney as research fellow and Christopher (Chris) Earl as research assistant.[148] Before joining the Open University in 1978, Rooney had studied mathematics at the University of Sussex and earned a PhD in applied mechanics from the University of Liverpool.[149] Earl joined the Centre for Configurational Studies in 1976 after completing a bachelor in mathematics and a master's degree with specialization in general relativity at the University of Oxford.[150] At the start of the 1980s, Rooney and Earl developed a popular third-level undergraduate course, Graphs, Networks and Design, prepared jointly by the Faculties of Technology and Mathematics and first offered in 1981.

The Science Research Council-funded project on architectural room layout planning developed ideas that March and Earl presented in a 1977 paper in *Environment and Planning B*. The paper's title, "On Counting Architectural Plans," may sound modest and tentative at first

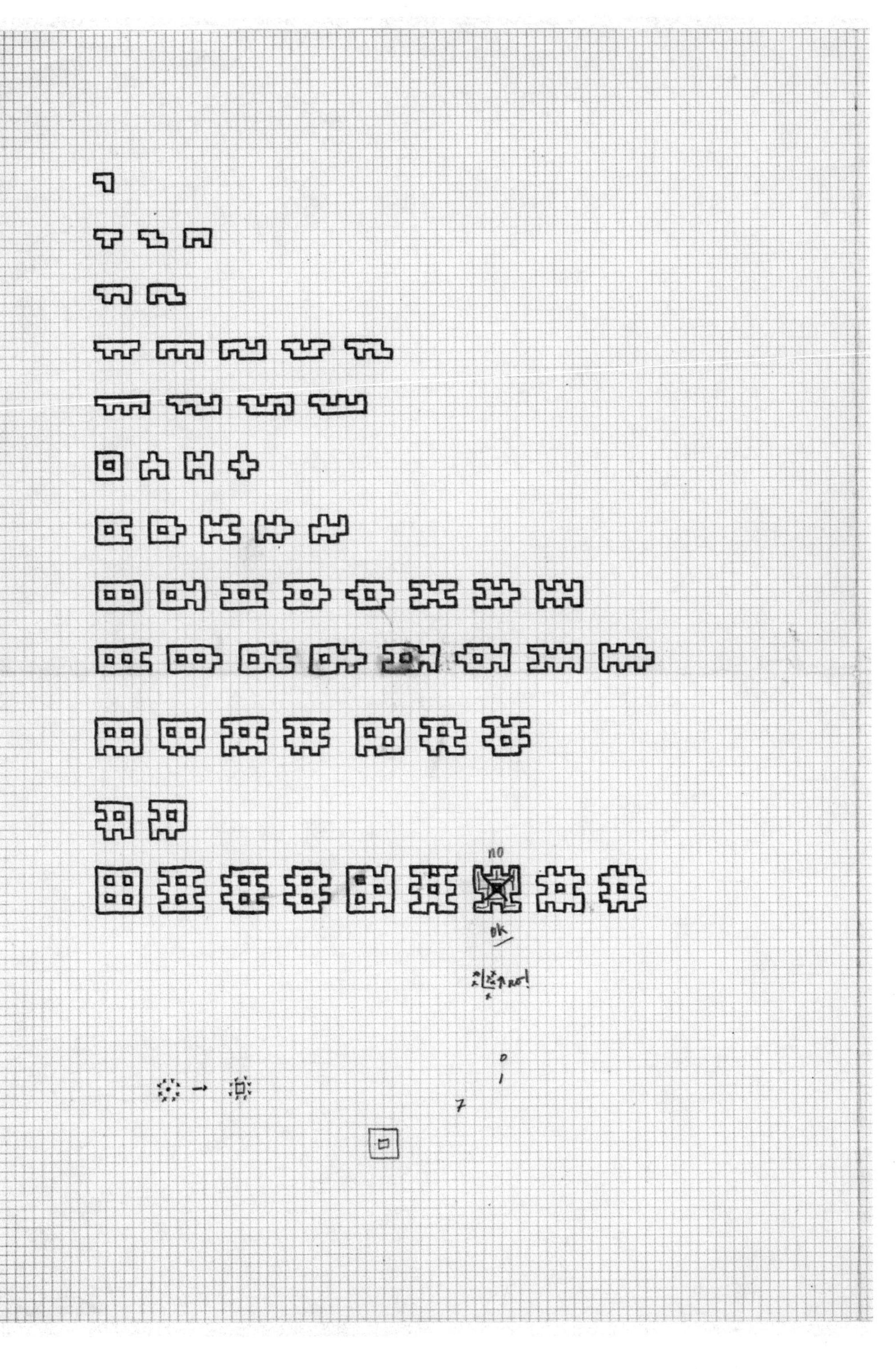

Figure 4.12
Sketches on graph paper of polyominoes with 2 to 10 faces. Lionel March, research on geometry, 1980, sketches in ink on paper, 42 × 59.2 cm. *Source*: Lionel March Fonds, Canadian Centre for Architecture, ARCH287269. Courtesy of the Lionel March Fonds, Canadian Centre for Architecture, Gift of Candida March. © Estate of Lionel March.

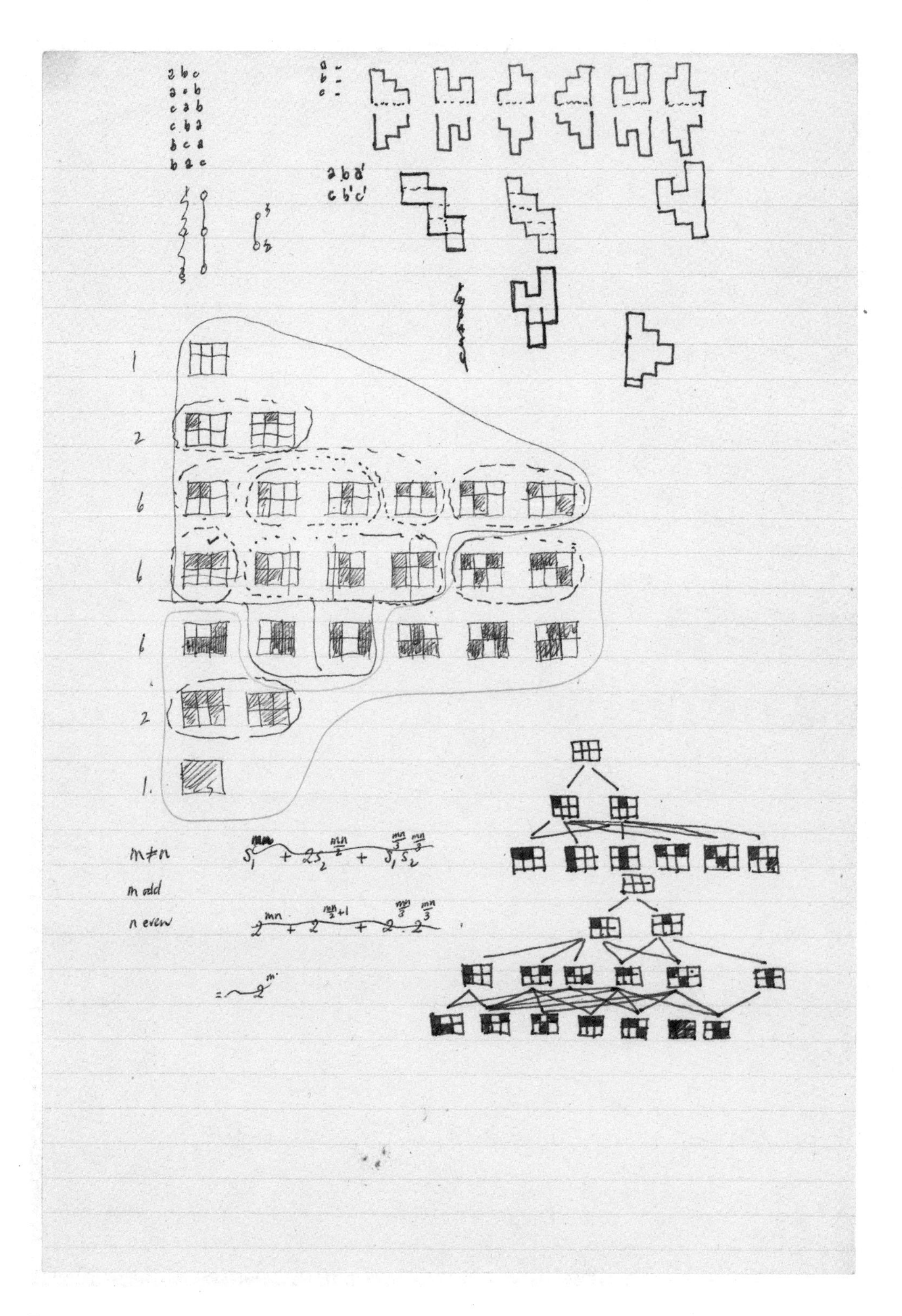

Figure 4.13
Pencil sketches of generations of polyominoes. Lionel March, research on geometry, 1980, sketches in ink and colored ink on paper, 29.8 × 20.8 cm. *Source*: Lionel March Fonds, Canadian Centre for Architecture, ARCH287270. Courtesy of the Lionel March Fonds, Canadian Centre for Architecture, Gift of Candida March. © Estate of Lionel March.

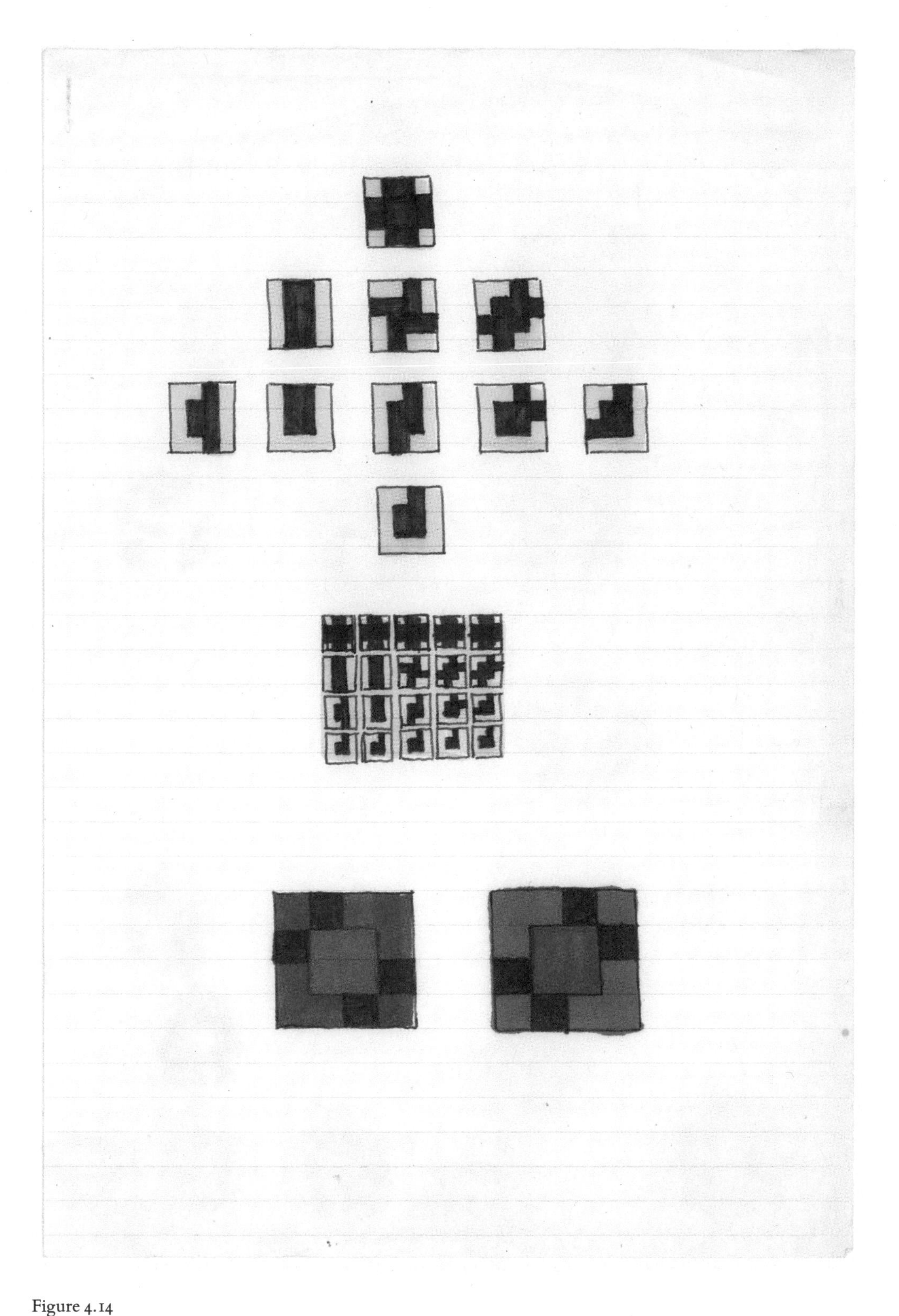

Figure 4.14
March's sketches of two-color polyominoes in red, yellow, and black ink. Lionel March, research on geometry, 1980, sketches in ink and colored ink on paper. *Source*: Lionel March Fonds, Canadian Centre for Architecture, ARCH287272. Courtesy of the Lionel March Fonds, Canadian Centre for Architecture, Gift of Candida March.

Interlude 8

Theories need their mythologies.[151]

The winged Melancholy recedes into the background as the sphere and the polyhedron propel forth.

The engraving is but a moment in the grand lineage of polyhedrists.

Pythagoras sowing the seeds for the study of polygons and polyhedra.

Vitruvius and Alberti reconnoitering polygonal form.

Dürer etching a polyhedral construction from a sphere.

Euler's mathematical study systematizing polyhedra.

Full circle to the hero of March's hero: Friedrich Froebel casting polyhedra as representations of harmony and order to shape, as March knew well, Wright's conception of "organic" architecture. "Organic" not as a metaphor for nature but as constructive patterns that had qualities of order and growth.

Froebel had after all worked as a curator of mineralogy in Berlin under crystallographer Christian Samuel Weiss.

Now graphs and polyhedra, together, would deliver a science of architectural form, once and for all.

Figure 4.15
Albrecht Dürer, *Melencolia I* (1514), detail. *Source*: The Metropolitan Museum of Art.

glance, but the paper's aspirations were the opposite. The goal was to give a definitive answer to the question of architectural possibility. To know what can be.

March and Earl set forth what they called a "Clopet demonstration" for architecture.[152] The phrase was taken from Louis Sullivan's autobiography, where the American architect reported being taken by his mathematics teacher's—Mr. Clopet's—invitation to throw a large book on descriptive geometry in the garbage. Every rule in the book had exceptions—a common complaint in diatribes against "old" geometry. Clopet's demonstrations, though, would be so broad as to eliminate exceptions. "If this can be done in Mathematics," March and Earl quoted Sullivan pondering, "why not in Architecture? The instant answer: It can, and it shall be!"[153] Instead of the "platitudinous slogan" (*form follows function*) that Sullivan, mentor of March's architectural hero Frank Lloyd Wright, thought to be the broad rule, the authors would offer a precise theory grounded in mathematics and science.[154]

Their "Clopet demonstration" for architecture launched from three objects in Albrecht Dürer's famous engraving *Melencolia I*, which March and Earl identified as the sphere, the carpenter's plane, and the polyhedron. These three elements were intertwined in a relationship of *construction*:

> Take up Durer's wooden sphere and carpenter's plane. From the sphere, plane off a simple flat section. A circular region is created. An enclosure is made. Architecture begins! Call the area contained by the circle a *room*, and the rest of the sphere *outside*. The circle itself is a mark of distinction, a *wall*.[155]

March and Earl continued to walk the reader through a process of cutting more facets of the sphere to add new rooms. Provisos applied. The facets should not be separate and should touch in such a way that no exterior comes between them. Each of these sphere configurations corresponded to an adjacency graph with a point representing each enclosure and a line representing a partition between the enclosures.[156] Using this principle, March and Earl defined the important notion of a "fundamental" architectural plan.[157] This was a plan in which all rooms (the points of the graph) connected to each other and each to the exterior with an edge (a line). March and Earl used terminology from Harary's 1969 *Graph Theory* to indicate that these graphs were "maximally planar," meaning no other edge could be added without making them nonplanar (that is, without making their edges intersect).[158] This also meant that all the areas between the edges (the graph's "faces") were triangles. From this condition stemmed a revealing assertion: no other architectural plan could be "more adjacent" than a fundamental plan.

Other plans could be derived from the fundamental plan through a process that March and Earl baptized with a remarkable term: *ornamentation*, taken from Wright's analogy of ornament to the blooming of a

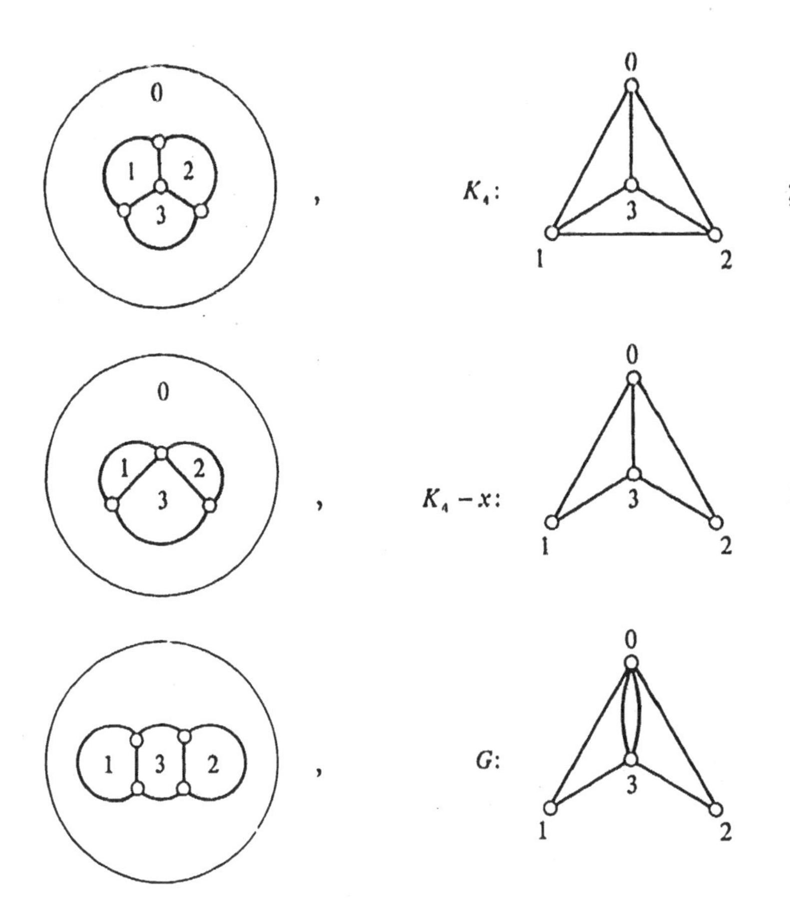

Figure 4.16
Cutting a sphere with planes. Every cut is an "inside," an architectural enclosure, and the sphere is the "outside." The graph represents the partitions between the enclosures. The graph's dual is the adjacency graph representing spatial structure. In this figure, the graph K4 is the fundamental architectural plan. *Source*: Lionel March and Chris Earl, "On Counting Architectural Plans," *Environment and Planning B* 4 (1977): 61. © 1977 SAGE.

flower or a tree in his 1953 *The Future of Architecture*.[159] The term nuances the story of graph vision because, although ornamentation suggested the existence of an invariant and preexisting structure, it did not operate on the surface of appearance. March and Earl construed ornamentation as "the material expression and enhancement of immanent, or underlying, structure."[160] Ornamentation consisted of both geometric interpretations and structural operations on the fundamental plans, such as deleting or adding partitions (edges on the maximal planar graph). It was what could be designed, the realm of choice, operating on a fixed and knowable infrastructure. They announced:

> At root we are saying that room formation is not itself a design problem, whereas ornamentation is. Immanent structures for each and every room formation are finite in number and are known aprioristically: architectural design is preeminently a matter of selection and the appropriate physical and material *transformation* [emphasis in the original] of one of these fundamental plans.[161]

The paper's demonstrations, following the "Waterloo school of combinatorics," performed a series of stunning jumps between the

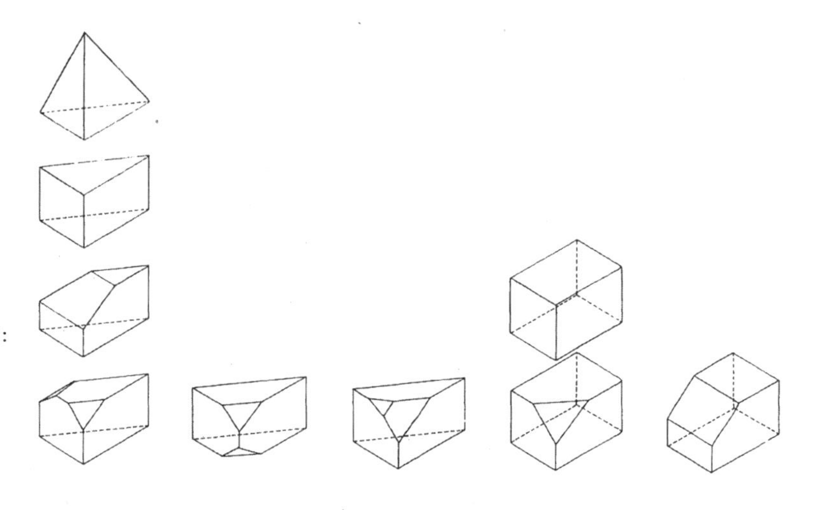

Figure 4.17
Trivalent 3-polytopes with up to seven faces. The first polyhedron gives "plastic expression" to the fundamental plan K4. The polyhedra are an alternative representation of fundamental architectural plans. *Source*: Lionel March and Chris Earl, "On Counting Architectural Plans," *Environment and Planning B* 4 (1977): 65. © 1977 SAGE.

triangular faces of maximal planar graphs and the faces of 3-polytopes (three-dimensional polyhedra).[162] Because their vertices and edges mapped one to one, polyhedra gave the graphs "plastic expression," and the graphs, in turn, were "skeletons" for the polyhedra.[163] The trivalent polyhedra (polyhedra with three edges meeting at each vertex), because of this mapping, represented fundamental architectural floor plans. Because of their triangular faces, maximal graphs could also be embedded as triangulations of the sphere. A stereographic projection (perspective projection of the sphere onto the plane) of these triangulations would give maximal planar graphs, which were the adjacency structures for the fundamental architectural plans.

Finally, March and Earl argued, every fundamental architectural plan given through this process could be drawn in the plane, orthogonally, so that every room was a polyomino. The paper's concluding remarks cautioned that "free forms" were "no freer than the right-angle" because the topologies of all fundamental plans could be accommodated in an orthogonal design.[164] The choice was a "question of design."[165] No additional freedom was to be gained by using curves instead of straight lines to represent graphs either—how to draw graphs was a "question of taste."[166]

These astonishing transpositions between spheres, graphs, polyhedra, and floor plans had acute consequences for architecture. The number of fundamental architectural plans was *known*; it sufficed to count trivalent 3-polytopes. The number of realizable architectural plans with a specific number of rooms was significantly lower than the number of possible adjacency requirements. Beginning with the adjacency graph was a mistake with dire computational consequences. This flipped user-oriented design processes on their head. "Do not ask a client what he wants," March and Earl instructed, "tell him what

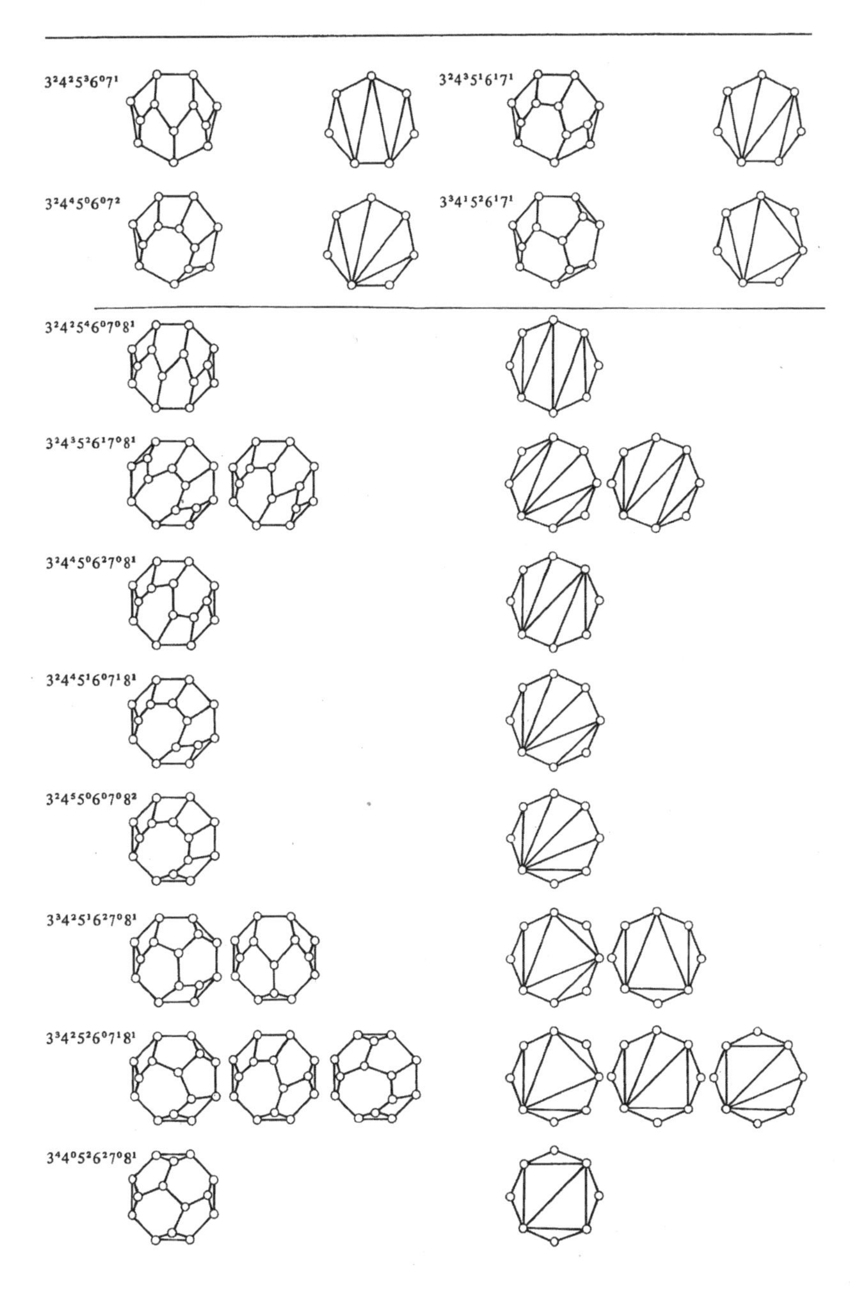

Figure 4.18
Fundamental architectural plans with seven and eight rooms.
Source: Lionel March and Chris Earl, "On Counting Architectural Plans," *Environment and Planning B* 4 (1977): 74. © 1977 SAGE.

he can have—it will be more helpful."[167] Freedom of choice required intimate awareness of design's skeletal infrastructures. So suggested the catechism: "Freedom in ignorance is false freedom: the only true freedoms in design lie through understanding and knowledge of the immanent structures upon which all designs must be based."[168]

Interlude 9

Readers of the third volume of *Environment and Planning B* in 1976 were presented with two exercises in formal composition.[169] Both were premised on the "beguilingly simple and architecturally interesting" endeavor of arranging spatial elements in an ordered way, the kind of pursuit that historian Paul Frankl recounted Leonardo puzzling with.[170] The exercise was to begin with a simple form, say a Greek cross, and then systematically arrive at all related central plan churches from that basic schema, by replacing its components with other shapes.

But what's in a cross?

The excerpt from Frankl suggested a 5-omino: "four square arms added to the sides of a central square."[171]

Elemental units in their unrelenting discreteness, the polyomino cells would aggregate, combine, and recombine.

But one could look at it differently.

The cross, readers were shown, could be broken down to 20 polygons.

These polygons were 20 distinct two-dimensional shapes.

With requisite rigor, readers were offered a definition of "shape" as a "finite arrangement of lines" that could "be drawn in a finite amount of time, for example on a piece of paper with a limited number of pencil strokes."

Shapes could be bound together with "spatial relations." One could define those with set theory, or simply draw them.

Classes of shapes could then be formed through shape rules that transformed a shape into another shape, mapping shapes onto shapes to generate languages of shapes.

What one saw in the cross suggested different rules forward.

The surface of appearance affected the computation.

An alternative foundation to March and Matela's plea for a "science of form" denounced combinatorics and stable decompositions, those "fixed collection of primitives out of which all shapes must be tediously specified."[172]

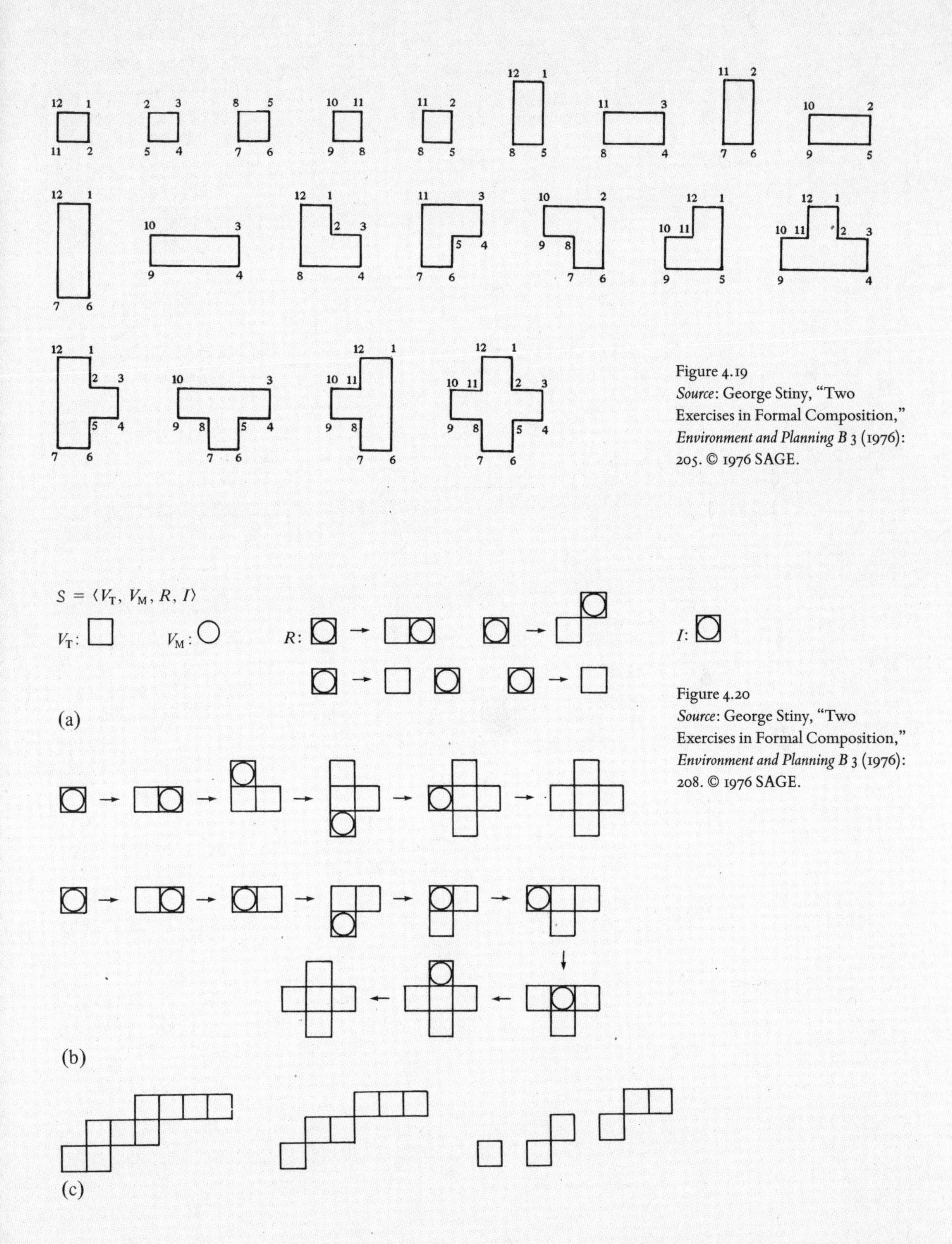

Figure 4.19
Source: George Stiny, "Two Exercises in Formal Composition," *Environment and Planning B* 3 (1976): 205. © 1976 SAGE.

Figure 4.20
Source: George Stiny, "Two Exercises in Formal Composition," *Environment and Planning B* 3 (1976): 208. © 1976 SAGE.

In 1977 George Stiny, trained in mathematics and computer science at MIT and having completed a PhD in the UCLA Systems Science Department under linguist, mathematician, and computer scientist Sheila Greibach, joined the Open University as a lecturer in design. There Stiny worked with Ramesh Krishnamurti—an electrical engineer trained at the University of Madras who studied and taught at the University of Waterloo between 1974 and 1975—on a two-year project on computer implementations of "shape grammars" and their applications to architecture funded by the Science Research Council.[173] The language of the grant's abstract, as presented in the Centre for Configurational Studies' first triennial report, was familiar—"automatic description, analysis, synthesis and evaluation of standard architectural objects."[174] Using shape grammars, Krishnamurti developed an algorithm acronymed "RK" for counting rectangular dissections, which he presented together with the "CB" algorithm by Cecil Bloch from the Martin Centre for Architectural and Urban Studies (the successor of the LUBFS Centre) in 1978.[175] Despite their mathematical and algorithmic commitments, shape grammars stood in juxtaposition to the Centre's other work for what Stiny would articulate as a resolute commitment to the surface of appearance.

Shape grammars were developed by Stiny and computer scientist James Gips as a redux of MIT linguist Noam Chomsky's transformational grammars.[176] Transformational grammars were sets of replacement rules indicated by arrows that were used to generate sentences by identifying types of linguistic elements in a phrase (noun, verbs, adverbs, . . .) and replacing them with other types of single elements or compounds of elements. Shape grammars adopted the idea of transformation rules but replaced linguistic elements (words) with spatial entities (shapes). In shape grammars, transformation rules took the form of A → B, where A, B were *drawings* of shapes (as opposed to symbols standing for shapes) and the → was a replacement operation. The replacement operation found an instance or a transformed (scaled, rotated, reflected) copy of a shape A in a drawing and replaced it with a shape B. Stiny framed this as a shift from "identity," in which a rule requires recognizing a specific symbol, to "embedding," in which a rule requires perceiving and picking out a shape, possibly from within another shape.

The interest in the diversifying effects of appearance came from the theory's rooting in the formalization of art. In 1974, Stiny and Gipps presented a paper titled "Formalization of Analysis and Design in the Arts" in Basic Questions of Design Theory—a symposium that Harary keynoted with enthusiastic observations about graph theory's pervasiveness in design theories and methods.[177] The paper adopted the mathematical parlance of structures and systems that characterized most symposium presentations, but with a focus on what the authors called "an aesthetic viewpoint," which they defined as a set of "interpretative conventions and evaluative criteria."[178] The authors explained that these conventions were applicable both analytically and generatively—two processes that they construed as "symmetric."[179] Aesthetic viewpoints were varied and relentlessly subjective, as many as a work's creators and onlookers. Yet this variety of viewpoints, argued Stiny and Gips, did not preclude the possibility of their precise (mathematical) statement.[180] The intellectual commitment to subjective multiplicity advanced mathematical scrutiny not on mathematical formalisms that declared different things to be deep down the same, but on formalisms that made it possible to see the same thing in different ways.

In a series of papers subsequently published in *Environment and Planning B*, Stiny presented sets and Boolean algebras to define shape and its transformations in an algebraic manner yet maintaining its visual ambiguity and properties as the driver of the computation. Because they consisted of infinitely many parts, shapes resisted a predefined structural description. Therefore, transformation rules needed to apply directly on the shape that one saw, on the surface of appearance. To achieve this, shape grammars used transformation geometry—the staple of the British new math geometry that traced back to Klein's "antimodern" endeavor to safeguard concrete experience in the realm of mathematical abstraction. Transformation geometry allowed for algebraic descriptions of shapes while preserving their visual concreteness.

In a 1982 letter to the editor of *Environment and Planning B* titled "Spatial Relations and Grammars," Stiny distinguished between "set grammars," which consisted of rules that treated designs as "*symbolic* objects [emphasis in the original]," and shape grammars, which treated designs as "*spatial* objects [emphasis in the original]."[181] Set grammars used rules called "spatial relations" to position or remove discrete objects (for example building blocks) in ways specified by the grammar's spatial relations (rules).[182]

Stiny presented set grammars as a compromise between the spatial attitude of shape grammars and the symbolic requirements of computer implementations.[183] Operating through identity, set grammars approximated the structural and combinatorial approach of graph-based architectural theories. Yet set grammars were presented as a case of a broader approach that went beyond points (symbols) and relations. In shape grammars, the structurelessness of shapes, which we heard many of this story's actors berating, was reframed as an asset. It afforded ongoing perceptual restructuring of the elements that one was manipulating during design, as opposed to mechanical combination of fixed entities upon an immanent infrastructure. Shape grammars were graph vision's computational antagonist.

Combinatorics

To talk about infrastructures is to talk about structures *beneath*. Infrastructures capture the condition of graph vision: a structure *under* concrete architectural designs that sets the limits of architectural possibility. They are March's "immanent structures" lying under all designs. They are Friedman's universal infrastructure with its ambiguous oscillation between the materialization of a mathematical abstraction and the mathematical abstraction of a material system. They are *A Pattern Language*'s cascade, holding together the appetizing smorgasbord of recipes for harmonizing form and its social and environmental contexts. Graphs as infrastructures negotiated control and freedom. They structured choice.

These infrastructural functions cast graphs as inherently political, a status complicated by their authors' claims to neutrality, featurelessness, necessity. Promoted as enablers of emancipation from architectural paternalism, ignorance, or the unreliable whims of architectural taste, graphs scaffolded claims to empowerment. These demands were timely engagements with a quivering architectural professionalism, a questioning of the architect's legitimacy as a decision maker, as someone who imposes their preferences and values on occupants, clients, "future users." This rhetorical realignment took place on a striking continuity of techniques. Adjacency graphs and other industrial engineering concepts were marshaled to conceptualize not what is *best* but what is *possible*—graphs expanded from *tools* to *infrastructures*.

Infrastructures were ostensibly value-free. Their limits, their proponents argued, were not to control, oppress, or exclude, but necessary and unavoidable: the laws of mathematics, of gravity, of space, the inherent logic of a collection of observations about the "environment." To fulfill their political position infrastructures were inevitably posited as theories, as systems that laid out rules with no exceptions—March and Earl's "Clopet demonstrations." Those who channeled graph-fueled imaginations of architectural possibility were watchful against errors that turned limits into curtailment. Slack mathematics and configurations were missed. Friedman's machine, Steadman cautioned, was wrong. Steadman, Mitchell, and Liggett's computer program, Earl warned, also missed options.[184]

Despite efforts to find representations that could reach the untenable vision of *exhaustive enumeration*, of a science of architecture, *une architecture scientifique*, that would clearly present what is possible, an irksome question lurked. Was the computation of possibility immune to the subjectivities of architects or architecture's "users"? Friedman suggested that these subjectivities would be expressed by freely navigating all possible options and making a choice with knowledge of the consequences, parsed through the (admittedly reductive) concept of "effort." The Architecture Machine's computational rendition of Friedman's theory in YONA and associated work imagined the system's user as impacting the system and introducing their preferences and biases on the kinds of architectural configurations they would be presented with. Despite the Architecture Machine's long-standing efforts to cultivate trust in the computer as a nondistortive mirror of its user, Friedman remained unconvinced.

In the Architecture Machine's quest to make people "their own architects" through personalized computing machines, Friedman still saw traces of the "paternalist scheme" that he condemned in his writings.[185] By learning its users' "idiosyncrasies," to use one of Negroponte's favorite terms, and acting on their behalf, even with "paternal benevolence," the computer bastardized the unmediated choice for which Friedman had advocated in the FLATWRITER.[186] In introducing the "Computer-Aided Participatory Design" section of Negroponte's *Soft Architecture Machines*, Friedman presented his concerns as stemming not only from his personal moral attitude, as he called it, but mainly from the pragmatic reason that "the learning about the personality of the future user is less implementable than the learning about structural characteristics of the real world."[187] The "implementability" of learning about the "real world," he explained, was not because it was less complex than human personalities, but because it was "by definition—more 'structurable'."[188]

In pledging faith in the "structurability" of the "real world," Friedman was documenting his immersion in a universe of material and abstract structures. In the architectural work that he pursued individually and as part of experimental architectural groups, space frames turned physical space into structured units. At the architectural scale, prefabrication further enforced processes of structuring (combining) discrete building components. Personal relationships with mathematicians and cultural actors such as Harary and Hill enchanted the mathematical techniques for depicting and computing these processes. Friedman took technorational concepts from industrial engineering and elevated them into a "democratic" design method rooted in axiomatic theory of a transparent architectural science. He linked evaluation methods of circulation efficiency with the moral imperative of "warning" the user; he recast the combinatorial generation of layouts as a register of open-ended possibility. More crucially, however, and despite his iconoclastic statements, he strongly advocated for the graph's intuitiveness and accessibility and imbued it with humanizing and democratizing meanings. Used to generate all possibilities through different combinations of its points, the graph was the infrastructure of an exhaustive "menu" of layout options.

In describing the influence of Friedman's methods on the Architecture Machine's work, Negroponte cautioned readers about "the particularly French notion of a 'banque de données' or what he [Friedman] calls a 'repertoire'."[189] The problem with these approaches, Negroponte explained, was that "the offerings of a menu of solutions obviously cannot exceed the combinatorial product of the parts (which may be enormous)."[190] Negroponte argued that Friedman escaped the particular constraint tainting French researchers by including in his repertoire "topologies that do not have a metric"[191]—in other words, graphs. "It is the user's adding of this metric," he concluded, "that affords the limitless variety."[192] An offspring of the Architecture Machine's collaboration with Friedman, the YONA system brought the graph onto the computer screen. Instead of mapping a sketch as in computer programs such as HUNCH or SQUINT, the graph was presented *as* the sketch—a malleable, visual entity, accessible well outside the cycles of a mathematical elite. By featuring the graph on the screen, the YONA system revealed its internal mathematical representation to its users. This revealing of the computer's interior to the eye not only abided by Friedman's moral mandate of transparency, but also implicitly declared the graph an interface between humans and machines: an entity that spoke the language of both. An abstract mathematical object with tangibly concrete visual depictions, the graph fostered optimism that human engagement with computers need not be seen in terms of dehumanizing quantification and mechanization. The graph infrastructure was a medium. Choice emerged as the result of a *conversation*—in cybernetician Gordon Pask's term—between the human and the machine through the congenial medium of the graph.

The Centre for Configurational Studies' forays on possibility had a different flavor, insisting on mathematical lucidity before engaging computer implementations and setting aside questions of interactivity and mediation. This was a stable point of critique for Architecture Machine members, who consistently argued that machines ought to understand the meaning of human decisions in order to assist in making them. At the Centre for Configurational Studies, humanism came through the participation in the age-long Western project of mathematical transcendence. Polyominoes, rectangular dissections, triangulations of the sphere would bolster a mathematically founded science of architectural form and would show, once and for all, what is possible. The promise here too was freedom, yet a freedom not founded on the promise of abundant choice but on the elegant delineation of its limits. Nothing was creative, they argued, about spatial configuration. All fundamental plans were knowable and countable before any action took place. Design began with operations on top of these essential infrastructures derived from graph-based combinatorics, as *ornamentation*.

Despite, or maybe because of, this boisterous commitment to the a priori infrastructures of design, the Centre for Configurational Studies also fostered a vocal counterpoint through Stiny's work on shape grammars. There, appearance did not operate as ornamentation occurring on top of a fixed structure, but it *restructured*, defined new entities and relationships through the continual interpretation and embedding of shapes by the human performing the computation. The sequence of rules and the realm of possibility shifted in every act. It was "participation" in the form that British philosopher Owen Barfield, whom Stiny spoke about during one of our meetings when I was his PhD student at MIT, quasi-spiritually invoked in his project on *saving the appearances*: an immersion in perceptual surface as the locus of computation.

5 Skeletons

Ghosts

I speak of "digital architecture" and I float between the Scylla and Charybdis of being cursory or being outdated. The risk is well worth it, though, because while this coda could not possibly characterize architecture's digital "culture," "turn," era, age, epoch with any semblance of historical or analytical precision nor foretell its fast-changing futures, it can discern the absences that construct it: the things that have been lost and those that have been hidden.[1]

"Digital architecture," in its contested but deliberate enchantment as a quest for new architectural *forms* that culturally converse with information technologies, communication networks, and digital media, has patently been about the surface. Seductive geometric forms rendered as the "digital's" ever-alluring architectural imagery displaced interest from mathematical processes to computationally generated architectural objects—ones that at one moment rejoiced in the seamlessness of their smooth, continuous surfaces.[2] An irresistible veil, these surfaces made mathematical structures recede from view and their pertinence in architectural discourse dwindle and fade.

In its most mundane form too, the all-familiar condition of clicking buttons and typing commands in design software packages, digital architecture, or doing architecture digitally, entails the perpetual hiding of mathematical and computational processes behind the graphics-rich surface of the screen. As I type these words, the "generative AI revolution" is immersing architects in "prompt-making" and interpreting relentlessly flat images churned by deep learning models trained, on dubiously acquired datasets, to turn words into pictures. An enthusiasm that will, likely, soon be old, outrun by new developments in the galloping data-hoarding and energy-devouring machine learning industry.

The image-saturated digital universe in which architects now reside is haunted. Structures of discrete entities define the operational universe of architects who work with computer technologies. Behind the screen of computer programs for drafting drawings, modeling forms, and managing building information lie entities—symbolic abstractions of shapes and building elements—combined and recombined based on relationships defined by the designer. Deep learning neural networks, the backbone of generative AI systems that are sweeping the cultural imagination today, are, well, *networks*. Architecture's entanglements with digital technologies are sodden with structural constructs—or as mathematician Claude Berge insisted we correctly name them, *graphs*.[3] We continue to think and operate in a graph-infused architectural universe.

There are, of course, graphs and graphs. The way neural networks compute has little to do with the techniques unfolded in the pages of this book. The graphs of neural networks are procedural; they show layers of information manipulation and are further stripped of any referential qualities—they lack the diagrammatic qualities that made the graphs of the story I have been telling here so compelling to architects. But despite these important differences, one aspect of graph vision has remained consistent. Skeletons lurked and lurk behind architectural images' visual appearances and geometric expressions, severing the perceptual from the operational realm. What one can do next on a drawing, on an image, on a model depends not on what one sees, but on a computational structure hidden beneath: the epitome of graph vision.

Graph vision was architecture's distinctive vernacular within the grand project of disciplinary modernization—a project of disciplinary consolidation and interdisciplinary unification after the Second World War, intertwined with new institutional formations of education and research. Graphs entered architecture propelled by this intellectual project's thrust, to ultimately shape it. I mean this quite literally: the graph was both a visual symbol and a way of practicing structural abstraction that was, in itself, transformative. Graphs grounded large intellectual agendas in concrete and workable mathematical techniques, which then formed the "material" for making architectural theories.

Graphs' theoretical embeddings in architecture were multiple and often antithetical. We saw tools for optimizing floor plans becoming theories for architectural democratization; methods for organizing design-related data shaping into a cookbook of patterns: shifts that graphs' persistence exposed as conceptually and technically continuous—a haunting of sorts, where the graph's tenacity invokes ghosts of modernism and its agonizing quest for certainty.[4]

Flesh

As graphs recede into the background, as they transform from willing collaborators in beguiling disciplinary projects to unwitting technical backdrops of architecture's digital condition, all we are left with are skeletons. This may seem like an odd thing to bewail. Graphs, after all, were always skeletons. And it was their skeletal form that appealed the most to their disciples. Graphs were the bones of architectural bodies, both bodies of knowledge, to paraphrase Boulding's influential "Skeleton of Science," and material bodies, underpinning architecture's drawn and built concreteness. They were thin, sticklike, weightless, like the skeletal infrastructures that Friedman was designing. Their mathematical rigor and intellectual purity, which rendered them free from the restraints of material form, kindled declarations of revolution—to recall March, Dickens, and Echenique's self-conscious manifesto. Architecture needed to be stripped of its soft tissues, its ever-proliferating contingencies, for its skeletal structures to appear. This was no simple task; it required a penetrating sight akin to an X-ray, beams picturing only architecture's ossifications, the immanent structures that supported its profuse expressions.

But graphs were no bare bones. They had cultural and material flesh. In their protean manifestations and contexts of use they were invariably embodied as material and cultural *things* that needed to be named explicitly, looked at, and computed with. They were drawn in different ways and through different media; they were called different names; they moved between objects and concepts. Graphs on the Architecture Machine's touch-sensitive IMLAC PDS-1D; graphs on quad paper in Friedman's notes; graphs in books; in research reports; scribbles in appendices to handwritten letters. Graphs' flesh was key for their architectural career. Their visual presentation, for one, entangled mathematical abstraction with architectural references—think again of Friedman's infrastructures. Visual presentation also served rhetorical purposes—recall Alexander's graph topologies as an index to shifting theoretical commitments.

Graphs' visual flesh makes this story about allure also one about deception. The graph's visual presentation was only deceptively visual because it attributed no transformative power to the perceptual realm. No matter how a graph is drawn, it is always the same thing: a set of discrete entities and relations. Things that *looked* dissimilar *being* the same, after all, was the premise of graph vision. These structural similarities would not require further analysis but would be immediately graspable by the eye. Isomorphism—a term that we saw March and Steadman praise with conviction—was about immediate appreciation of underlying sameness. Mathematical comparison of graphs in an analytical manner was there to protect the eye from being deceived by potential differences in the graphs' visual presentation. Yet along with a belief in the unifying power of structural abstraction came a disregard for the diversifying effects of concrete appearance. It did not matter what the graph's points represented. The properties of things evaporated under the power of their relations.

Yet, taken together, graphs' embodiments put their finger on graph vision's ambivalence: calls for disembodied abstraction conflicted with an insatiable appetite for drawing, seeing, and making, a patent disregard for perceptual appearance in tension with a maniacal drawing of graphs. Ambivalence yields possibility. It is about negotiation, indecision, and bifurcation. Graphs' hiding, reification, and naturalization in all things "digital" causes their material and cultural flesh to decompose. The ambivalence is muted. Disembodied skeletons puppeteer the lush imagery architects produce and consume, proper infrastructures of possibility. It is easy to forget they are there.

Boxes

The cultural flesh of architecture's erstwhile revered skeletons is left to rot in boxes. A box is a fitting metaphor for a digital computer, even when its material form changes from a mammoth case to a thin silver slice. Here we stumble upon another object that structures metaphors around computer systems' opacity, the "black box." A black box—an idea birthed from a material artifact carrying a radar component from British to Canadian and American allies during the Second World War—is something that can only be known through its inputs and outputs.[5] The internal workings remain hidden from view. "Blackboxing" is about obscurity, unknowability. Today, demands to "open the black box" together with calls for algorithmic "accountability" and "explainability" rise with urgency, confronting the scale and power of computing systems in personal, social, and political spheres. In architecture, concepts such as "gray-boxing" have recently attempted to reconcile the growing opaqueness and encapsulation of data and computational commands with demands for transparency and control.[6]

The "black box" and its opposite, the "glass box," have historically functioned as poles around which to ponder architecture's relationship with mathematics and computing. A well-known example is architect and urban designer Jonathan Barnett's report on the first conference of the North American Design Methods Group held at MIT in 1968—an event that included, as we have seen, papers on URBAN5, environmental structure, and graph representations of floor plans, among others.[7] Talk of "glass boxes" and "black boxes" was familiar to the design methods community from J. C. Jones's book *Design Methods: Seeds of Human Futures*.[8] The terms pointed to different attitudes toward the knowability of the design process and the designer's cognitive machinations, the ability to render them fully and completely transparent. But the "glass box" and the "black box" also figured as two distinct containers for mathematical and algorithmic abstractions.

"Glass box" approaches orbited around a mathematical and logical approach to the design process, in the fashion of Alexander's early work on decomposition and recomposition. The design process was restructured. Computers were mere executers of algorithmic instructions, performing faster and more efficiently a process that was entirely transparent to a human operator. The computer program was a clear box, its internal workings in plain view. "Black box" approaches, mainly associated by Barnett with computer graphics research, accepted the indescribability of the designer's operations. They attempted to enhance partial tasks that designers faced in a traditional process with the aid of new graphical and interactive technologies. This apparent rift did not seem the least disconcerting to Barnett, who saw in it a "formidable combination" of "computer technology applied to design problems in a subtle way, plus more rigorous methods of using the intricate design abilities of the human mind," and advised architects to "keep an eye on the Design Methods Group."[9] If nothing else, Barnett explained, the conference indicated the existence of "a significant group of people, many of them still in their 20's, who possess a sufficient knowledge of both architecture and of the mathematics needed to work with computers that they can deal intelligently with either one."[10]

Graph vision was, for the most part, about glasslike transparency. It was about revealing, bringing out in the open, shining light on architecture's immanent structures. Implementations on digital computers mattered, they pushed for ever better and more efficient algorithms, but there was also computation without computers: a Xeroxable book of patterns as a generative architectural method, pages on pages of hand-drawn permutations of graphs representing "fundamental plans." The computer as a black box was often seen as inimical to the project of graph vision. Recall Friedman's concerns about the Architecture Machine's participatory systems' paternalism: the FLATWRITER's politics were tied to the system's transparency and mechanical combinatorics—what Steadman derided as the proverbial monkeys typing Shakespeare.

And then think of the Architecture Machine and its conversational interfaces, keen on cultivating the illusion that the machine is not a dry mechanical calculator but a congenial partner. These tendencies became all the more entrenched by the group's collaboration with cyberneticians Gordon Pask and Warren Brodey, both leaders in the field that turned the "black box" from a material thing to a structuring metaphor for a new knowledge paradigm that replaced the relentless representational illumination of post-Enlightenment science with the performative utterances of things that are obscure and unknowable.[11] The Architecture Machine hid the computer's workings from its users. And yet graph-enabled isomorphisms between a cognitive structure of user needs and desires, their graphic utterance as a sketch hiding a spatial structure, and the graph mapping of this structure in the computer professed to ensure an operational transparency. The isomorphic mapping of sketches' inferred structure onto graphs claimed undistorted translations between the graphical statements and mathematical structures with which the machine could compute. The Architecture Machine's computer-aided participatory architecture system purported to map between graphical descriptions, structures of intentions, and structures of physical space. The graph as a common representation of all three domains was paramount in supporting claims of conversational fluidity with the computer. The black box was to be trusted; it was, after all, Your Own Native Architect.

Closets

But maybe we shouldn't talk about boxes.

We should talk about closets.

A closet is a favorite site of secrecy, of concealment. It is a container of stories, of forgotten pasts and repressed desires. A closet is intimate; it is private. It is a space. A closet can open and close. Things can be put in and taken out. One can get in and come out. Closets are also portals to other worlds. Their suffocating spatial containment is almost implausible.[12]

An architectural enclosure, the closet promises a disciplinary inflection to the contemplation of mathematical and algorithmic containers. Unlike boxes, which echo military expeditions, critical secrets, and networks of control, closets are domestic and personal. They speak to architects' work in domesticating techniques from military and industrial operations and embedding them into culture. It is the kind of formidable, if thorny, transposition that takes a calculation of workers' movement on the factory floor and recasts it as an avenue toward architectural participation and self-actualization. In giving mathematical and algorithmic techniques cultural form, architects—not unlike computer artists—contemplated distinct but ever-inventive ways that algorithms and computation could be entangled with the grain of everyday life. There was also another kind of domestication at play: finding an architectural home for new and exciting mathematical varieties that for many promised no less than a twentieth-century architectural renaissance.

Closets might also be about illicit knowledge. The mathematical bodies in architects' closets or the practices of acquiring them are nowhere close to the "skeletons in the closet" saying's ominous, if fabricated, provenance. And yet the tale attunes us to crossing boundaries, to working in the margins, to access; to the impetus to continue opening architectural bodies up, under conditions of scarcity: scarcity of data, of computing resources, of mathematical knowledge.

In the closet lies the memorabilia box. Purposely tucked away. In it, a textbook from school mathematics, several pages of cell-like jottings, a heated debate with a mathematical pen pal, doodles of pattern topologies—biographical snippets in the making of a collective graph vision.

Do not mistake this story's overwhelmingly male cast of characters as a celebration of great men. They are not this story's protagonists. They are carriers of graphs' sweeping movement through architecture, a movement full of shape-shifting and disguise. Architects' graphs queer abstract mathematical skeletons by activating the forces that graphs were enlisted to suppress—visual ambiguity, serendipitous similitudes, the seductiveness of appearance. Graphs were there to be looked at all along, troubling iconoclastic declarations with forbidden indulgence in seeing and drawing—closeted desires of architecture's computational imagination in which images merely follow aniconic calculations. Graph vision is the story of that ambivalence toward granting the surface and shapes operative power. Mathematics and computation, as we have seen, do not need to signal dematerializing abstraction and disembodied skeletons. But first we need to make amends with shape. Surfaces are profound. There is no shame in being superficial. Superficial, as the philosopher and poet wanted it, out of profundity.[13]

Acknowledgments

This book has come out of love: love for mathematics, for drawing, for writing, for telling stories, for meandering through archives, for building worlds, and for the thrill of traversing them. It has also come from generosity: the generosity of mentors, colleagues, archivists, students, and friends who gave ideas, time, energy, support, and intellectual fuel to this book throughout the years of its making. As this project leaves palimpsests of drafts, and folders over folders of images, I am overcome by gratitude and appreciation for the community that made telling the story of graph vision a joyful and collective journey.

The making of this book started during my PhD studies in design and computation at the Massachusetts Institute of Technology. I am indebted to my PhD supervisor and mentor George Stiny for his profound commitment to computation as a theoretical, philosophical, and cultural pursuit. The sessions he led on shape grammars in our doctoral "salon" opened exciting avenues toward theories of perception and seeing, pragmatist philosophy, romantic poetry, and the critiques of binary and discrete computation. Through commitment to the transformative power of seeing, shape grammars offered a counterpoint, a ground, for me to recognize what I call "graph vision" as a distinct culture in architectural computing, to seek out its histories, and to contend with its limitations.

Deep gratitude for helping shape this project's theoretical and methodological armature also goes to Terry Knight, who offered continuous support and encouragement to my explorations in computation and its history, before and after my time at MIT. Her research and teaching on feminist approaches to computing buttressed the critical sensibilities of this book and cultivated my attention to the analytical significance of material practices, even when those who perform them deny their existence.

I also wish to warmly acknowledge Natasha Schüll and Timothy Hyde, both erstwhile teachers of mine at MIT and members of my doctoral dissertation committee, for offering me a rich gamut of methods in the history of science and technology and in the history of architecture, and for being sharp and thoughtful readers my work.

This book bears the marks of inspiring mentors who are no longer here: the late Patrick Winston, whose heuristics on communication and forays in cultures of artificial intelligence—alongside his kindness and personal support—will stay with me forever; the late Edith Ackermann, one of the brightest and most inspiring voices in our doctoral "salon" at MIT; and the late Dimitris Papalexopoulos, who introduced me to the fascinating world of the philosophy of design and technology while I was still a graduate student in Athens.

I also warmly thank my professors at MIT, Arindam Dutta, Leah Buechley, Caroline Jones, Jennifer Light, Clapperton Mavhunga, David Mindell, Hannah Rose Shell, and Sherry Turkle, for the tools they gave me to approach the entanglements of science, technology, architecture, and art.

I am grateful to colleagues at MIT with whom I began thinking about this project and who have continued to offer community and support: Moa Carlsson and Athina Papadopoulou for their friendship and endless conversations and, in alphabetical order, Masoud Akbarzadeh, Asli Arpak, Alexandros Charidis, Irina Chernyakova, Felecia Davis, Antonio Furgiuele, Onur Yüce Gün, Sarah Hovsepian, Josh Ingram, Moritz Kassner, Devin Kennedy, Clare Kim, Duks Koschitz, Sotirios Kotsiopoulos, João Magalhães Rocha, Laia Mogas-Soldevida, Vernelle Noel, Mine Özkar, Dimitis Papanikolaou, Will Patera, Diego Pinochet, Daniel Rosenberg, Carolina Soto-Ogueta, Alan Song-Ching Tai, Alise Upitis, Katia Zolotovsky for helping me shape the foundations of this project through our conversations then, and since then.

If my time at MIT lay the seeds of this book, my current academic home, the School of Architecture at McGill University, is where this book has grown and flourished. I am indebted to Martin Bressani for the indispensable encouragement and support in making this publication a reality. His advice on academic publishing and feedback on the prospectus and chapters of this book have been invaluable. I am also deeply grateful to David Theodore for thinking with me about graphs, string diagrams, and algorithmic automation, and for always making time for debate, advice, and open-ended conversation. I also cordially acknowledge the material support he has offered this book as the Director of the School of Architecture. Heartfelt gratitude goes to Ipek Türeli for the comradery and guidance as I have been navigating contexts of research, work, and life in Montreal. I am thankful to Annmarie Adams for the enriching conversations on historical methods and for motivating detective work on important historical details. Sincere appreciation extends to past and current colleagues at the School of Architecture. Warm acknowledgments extend to the school's administrative officer David Krawitz and to the media technician Juan Osorio for assisting with various practical issues around the production of the book.

I am grateful to my doctoral supervisees Eliza Pertigkiozoglou and Maxime Leblanc for the lively intellectual exchange and multifaceted hands-on support, to Nathalie Kerschen for discussions on formalisms and research collaborations, to Cigdem Talu for her diligent work with securing permissions for the large collection of images featured in the book, and to all my students at the School of Architecture for their engagement with key ideas and concepts of this book in the context of my seminars, lectures, and research projects.

This book has been growing alongside transformative exchanges with an expansive group of colleagues and friends near and far whom I am fortunate to have in my intellectual community: in alphabetical order, Matthew Allen, Peder Anker, Daniel Barber, Henriette Bier, Roberto Botazzi, Ewan Branda, Nathalie Bredella, Anne-Marie Brennan, Sina Brueckner-Amin, Craig Buckley, Ultan Byrne, Galo Canizares-Proano, Mario Carpo, Zeynep Çelik Alexander, Gerardo Con Diaz, Chris Dähne, Martien de Vletter, Stephanie Dick, Shelby Doyle, Teressa Fankhänel, Yuriko Furuhata, Jacob Gaboury, Nadja Gaudilliere-Jami, Bernard Geoghegan, Rania Ghosn, Tom Haigh, Orit Halpern, Mark Hayward, Lily Irani, Konstantina Kalfa, Lydia Kallipoliti, Hanan Kataw, Evangelos Kotsioris, Adam Lauder, Areti Markopoulou, Shannon Mattern, John May, Anna-Maria Meister, Constantinos Miltiades, Ajung Moon, Philippe Morel, Michael Osman, Elizabeth Patitsas, Christopher Phillips, Antoine Picon, Rachel Plotnick, Dennis Pohl, Cailen Pybus, Angie Rout, Rafico Ruiz, Akshita Shivakumar, Ranjodh Singh Dhaliwal, Molly Steenson, Alma Steingart, Rebecca Uchill, Shota Vashakmadze, Kathy Velikov, Georg Vrachliotis, and Andrew Witt.

I am indebted to Daniel Cardoso Llach for motivating me through his rigor, kindness, and impeccable scholarship. Working with him in curatorial and editorial projects has been a process of continual learning and growth. I thank him for being a friend, a supportive colleague, and a tireless interlocutor in conversations about sociotechnical systems and cultural histories of computing.

A big, hearty thank you goes to Olga Touloumi for the long and unconditional friendship, her ever-attentive ears and forthright advice, and for helping me grow as a scholar. Her incisive criticism, ethical compass, and research acumen have been irreplaceable. I am grateful to her for our discussions on a polyglot space for architectural computing and her dedication to making that space happen through our organizational and editorial collaborations.

I warmly acknowledge Michelle Niemann for her feedback, encouragement, and infectious positivity throughout the writing of this book.

This book would have not been possible without the invaluable support of some of its primary actors and their archivists who gave me access to their stories and materials. I am grateful to Philip Steadman for sharing rich stories of his mathematical education, openly discussing his early work with me, and generously allowing me to feature material from his publications in this book. I was fortunate to have had the chance to speak with the late Yona Friedman and the late Lionel March before their passing, and more recently to have had access to their newly established archives. A most cordial thank you goes to Marianne Friedman-Polonsky, to Jean-Baptiste Decavèle, and to the Fond Denise et Yona Friedman for making Yona Friedman's archive publicly available and giving me permission to use some of the incredibly rich material it hosts. Warm thanks go to Martien de Vletter, Anna Haywood, and Chloe Belair-Morin at the Canadian Centre for Architecture who gave me early access to the Lionel March Fonds. I am most grateful to Maggie Moore Alexander, Artemis Anninou, and the Christopher Alexander / Center for Environmental Structure Archives for so openly sharing resources and material. I also wish to gratefully acknowledge Christopher Earl, Marcial Echenique, and Guy Weinzapfel for offering me permissions to reuse their work.

A resounding thank you goes to Thomas Weaver, whose care and appreciation for the craft of writing, elegance of structure, and the book as a material and aesthetic form have been vital for the development of *Graph Vision*. I thank him for a wonderful collaboration full of energy, wit, and insight. I am deeply appreciative of Matthew Abbate's impeccable work copy-editing this book and to Yasuyo Iguchi and Jay Martsi for helping me tell the visual story that is the story of graph vision through the book's graphic design. I thank Tobiah Waldron for his work on the index.

I am wholeheartedly grateful to my father Dimitris for infusing me with his love for mathematics and logic and to my mother Fani for teaching me to draw, paint, and learn to see. My endless thanks go to my sister Rodanthi for being my closest friend, my role model, one of the best writers and sharpest thinkers I know, and for reading drafts on drafts of the manuscript.

All my love and appreciation to my partner Niko for his support, unfaltering encouragement, and companionship. *Graph Vision* is dedicated to our children, Aliki and Dimitri, my bright sparks and biggest inspirations.

Notes

Chapter 1

1. Blanche Descartes, "The Expanding Unicurse," in *Proof Techniques in Graph Theory*, edited by Frank Harary (New York: Academic Press, 1969), 25.
2. Rob Shields, "Cultural Topology: The Seven Bridges of Königsberg 1736," *Theory, Culture and Society* 29, no. 4–5 (2012): 43–57.
3. Joseph Malkevitch, "Pseudonyms in Mathematics," American Mathematical Society, 2020, http://www.ams.org/publicoutreach/feature-column/fc-2020-12
4. Cedric A. B. Smith and Steve Abbott, "The Story of Blanche Descartes," *Mathematical Gazette* 87, no. 508 (2003): 23–33.
5. Gilles Retsin, ed., "Discrete: Reappraising the Digital in Architecture," *Architectural Design*, no. 258 (2019).
6. Arjun Appadurai termed "methodological fetishism" an approach that centers on *things* as entities that can be "followed" across the social contexts that enliven them. See Arjun Appadurai, ed., *The Social Life of Things: Commodities in Cultural Perspective* (Cambridge: Cambridge University Press, 1986), 5.
7. Bill Brown rebuts criticisms of "methodological fetishism" as disavowing psychological and phenomenological approaches to the production of materiality. He argues that methodological fetishism makes it possible to think of "how inanimate objects constitute human subjects, how they move them, how they threaten them, how they facilitate or threaten their relation to other subjects." Bill Brown, "Thing Theory," *Critical Inquiry* 28, no. 1 (2001): 7.
8. Hans-Jörg Rheinberger, *Toward a History of Epistemic Things: Synthesizing Proteins in the Test Tube* (Stanford, CA: Stanford University Press, 1997), 1. Lorraine Daston grants scientific objects at once a *real* and a *historical* status and argues for the revival of ontology for historians of science. This motivates the crafting of "biographies of scientific objects," as she terms them. In his history of mathematics and architecture in the twentieth century, Andrew Witt has adapted this notion to speak of "biographies of methods." Also of relevance here is Ian Hacking's concept of "historical ontology," which, though originally applied to psychology, language, and thought, pays attention to the conditions that make existence possible. See Lorraine Daston, "Introduction: The Coming into Being of Scientific Objects," in Daston, ed., *Biographies of Scientific Objects* (Chicago: University of Chicago Press, 2000), 14; Andrew Witt, *Formulations: Architecture, Mathematics, Culture* (Cambridge,

MA: MIT Press, 2022); Ian Hacking, *Historical Ontology* (Cambridge, MA: Harvard University Press, 2022).

9 The term "knowledge setting" is taken from Karin Knorr Cetina's landmark work on epistemic cultures, which she has famously defined as "cultures that create and warrant knowledge." Knorr Cetina's work has championed a practice approach to knowledge, attuned to the specificities of *settings*—structures, processes, and environments in which knowledge is produced and sanctioned. See Karin Knorr Cetina, *Epistemic Cultures: How the Sciences Make Knowledge* (Cambridge, MA: Harvard University Press, 1999), 1, 8.

10 For a definition of "vision," see *Oxford English Dictionary*, s.v. "vision (n.)," February 2024, https://doi.org/10.1093/OED/6460679308. In the introduction to her monograph *Beautiful Data*, Orit Halpern discusses "vision" discourse as self-divisive and self-differentiating; as a discourse that "multiplies and divides from within." Orit Halpern, *Beautiful Data: A History of Vision and Reason since 1945* (Durham: Duke University Pres, 2015), 21.

11 Halpern, *Beautiful Data*, 22. On vision and visualization see also Johanna Drucker, *Graphesis: Visual Forms of Knowledge Production* (Cambridge, MA: Harvard University Press, 2014).

12 Erwin Panofsky, *Studies in Iconology: Humanistic Themes in the Art of the Renaissance* (1939; Boca Raton, FL: Taylor and Francis, 2018).

13 Hal Foster famously referred to "historical techniques of sight" and "discursive determinations" of vision as "visuality." Hal Foster, ed., *Vision and Visuality* (Seattle: Bay Press, 1988), ix.

14 Daniel Cardoso Llach uses "vision" to evoke a new "imagination" of design that emerged from early discourses on computer numerical control and computer-aided design, as well as reconfigurations of agency and labor within these systems. He suggests that these early experiments laid the intellectual and technical groundwork for designers' discursive and practical engagements with software today. Daniel Cardoso Llach, *Builders of the Vision: Software and the Imagination of Design* (New York: Routledge, 2015).

15 Jacob Gaboury foregrounds the analytical conundrum that stems from computer graphics' "phenomenal invisibility as a distinct technical medium." The medium of computer graphics, he argues, is chiefly discussed in terms of the ways that it produces its own invisibility, that it occludes technical processes behind a screen and an image. See Jacob Gaboury, *Image Objects: An Archaeology of Computer Graphics* (Cambridge, MA: MIT Press, 2021), 3, 4. On the function of a screen as a filter, as something that hides, see also Francesco Casetti and Bernard Dionysius Geoghegan, "Screen," in *Information: Keywords* edited by Michele Kennerly, Samuel Frederick, and Jonathan E. Abel (New York: Columbia University Press, 2021), 162–173.

16 R. E. Somol, "Dummy Text or the Diagrammatic Base of Contemporary Architecture," in *Diagram Diaries*, edited by Peter Eisenman (New York: Universe Publishing, 1999), 6–25. On debates around the architectural diagram see also Mark Garcia, ed., *The Diagrams of Architecture* (Chichester, UK: Wiley, 2010); and Anthony Vidler, "Diagrams of Diagrams: Architectural Abstraction and Modern Representation," *Representations* 72 (2000): 1–20.

17 For influential accounts of the digital avant-gardes of the long 1990s see, for example, Mario Carpo, *The Digital Turn in Architecture 1992–2012* (Chichester, UK: Wiley, 2013); Antoine Picon, *Digital Culture in Architecture: An Introduction for the Design Professions* (Basel: Birkhäuser, 2010); Greg Lynn, ed., *Archaeology of the Digital* (Montreal: Canadian Centre for Architecture, 2013).

18 Ben Van Berkel and Caroline Bos, eds., "Diagram Works: Data Mechanics for a Topological Age," *ANY Magazine* 23 (1998).

19 The generation of architects that succeeded the instigators of the modern movement in architecture used the term "heroic" to denote the revolutionary spirit of modern architectural theory in the 1920s and 1930s. The term appeared in a 1965 article by Alison and Peter Smithson, leading critics of Le Corbusier's or Sigfried Giedion's doctrinaire approach to urbanism within CIAM and founding members of Team 10. See Peter Smithson and Alison Smithson, "The Heroic Period of Modern

Architecture," *Architectural Design* 35, no. 12 (1965): 587–642, and their subsequent book *The Heroic Period of Modern Architecture* (New York: Rizzoli, 1981).

20 Sarah Williams Goldhagen and Réjean Legault, eds., *Anxious Modernisms: Experimentation in Postwar Architectural Culture* (Cambridge, MA: MIT Press; Montreal: Canadian Centre of Architecture, 2001).

21 George Kasabov, John Outram, Paul Power, and Ian McKechnie, "Student Section—BASA," *Architects' Journal* 129 (1959): 451.

22 On modern architectural theory's staying power see also Stanford Anderson, "The 'New Empiricism–Bay Region Axis': Kay Fisker and Postwar Debates on Functionalism, Regionalism, and Monumentality," *Journal of Architectural Education* 50, no. 3 (1997): 197.

23 Avigail Sachs uses the term "second modernism" to point to the American architects' resistance to modern aesthetic canons and pursuit of another kind of modern architecture founded on research. See Avigail Sachs, "Research for Architecture: Building a Discipline and Modernizing the Profession," dissertation, University of California, Berkeley, 2009. Arindam Dutta also uses the term in the title of his 2013 edited collection, which explores architecture's entanglements with technoscientific practices and ideals in the context of MIT: Arindam Dutta, ed., *A Second Modernism: MIT, Architecture, and the "Techno-Social" Moment* (Cambridge, MA: MIT Press, 2013). On the renegotiation of modernism and radical recalibrations of design pedagogy within architecture schools operating in research-oriented universities, see also essays in Joan Ockman, ed., *Architecture School: Three Centuries of Educating Architects in North America* (Cambridge, MA: MIT Press, 2012); and Beatriz Colomina, Ignacio G. Galán, Evangelos Kotsioris, and Anna-Maria Meister, *Radical Pedagogies* (Cambridge, MA: MIT Press, 2022).

24 Historians of science and higher education have attributed to "basic research" the status of an *ideology* in Cold War America. See, for example, Nathan Reingold, "Vannevar Bush's New Deal for Research: Or the Triumph of the Old Order," *Historical Studies in the Physical and Biological Sciences* 17 (1987): 299–344; Roger L. Geiger, *Research and Relevant Knowledge: American Research Universities since World War II* (New York: Oxford University Press, 1993).

25 Le Corbusier, for instance, famously promoted the *modulor*—a proportional system based on human measurements and mathematical ratios such as the golden section or the Fibonacci numbers—as humanizing the mechanical universe of industrial mass production. Le Corbusier wrote about mathematics as a form of metaphysics: "Mathematics is the majestic structure conceived by man to grant him comprehension of the universe" (*Le Modulor I* [Boulogne, 1950], 71). Or elsewhere, "For the artist, mathematics does not consist of the various branches of mathematics. It is not necessarily a matter of calculation but rather of the presence of a sovereign power; a law of infinite resonance, consonance, organisation. . . . Chance has no place in nature. Once one has understood what mathematics is—in the philosophical sense—thereafter one can discern it in all its works. Rigour, and exactness, are the means behind achieving solutions, the cause behind character, the rationale behind harmony" ("L'architecture et l'esprit mathématique," in *Les grands courants de la pensée mathématique* [Paris, 1948], 490). Both cited in Judi Loach, "Le Corbusier and the Creative Use of Mathematics," *British Journal for the History of Science* 31, no. 2 (1998): 185.

26 Jeremy Gray, *Plato's Ghost: The Modernist Transformation of Mathematics* (Princeton, NJ: Princeton University Press, 2008).

27 Nicolas Bourbaki, "The Architecture of Mathematics," *American Mathematical Monthly* 57, no. 4 (1950): 221.

28 Maurice Mashaal, *Bourbaki: A Secret Society of Mathematicians*, translated by Anna Pierrehumbert (Providence, RI: American Mathematical Society, 2006).

29 Alma Steingart, "The Axiomatic Aesthetic," in *Computer Architectures: Constructing the Common Ground*, edited by Theodora Vardouli and Olga Touloumi (London: Routledge, 2020), 194–213.

30 Ivor Grattan-Guinness, *The Search for Mathematical Roots, 1870–1940: Logics, Set Theories and the Foundations of Mathematics from Cantor through Russell to Gödel* (Princeton, NJ: Princeton University Press, 2011), 208.

31 David Aubin, "The Withering Immortality of Nicolas Bourbaki: A Cultural Connector at the Confluence of Mathematics, Structuralism, and the Oulipo in France," *Science in Context* 10, no. 2 (1997): 297–342.

32 Bernard Dionysius Geoghegan offers a rich critical account of structuralism's entanglements with technical regimes of information theory and cybernetics. See Bernard Dionysius Geoghegan, *Code: From Information Theory to French Theory* (Durham: Duke University Press, 2023); and the earlier Bernard Dionysius Geoghegan, "From Information Theory to French Theory: Jakobson, Lévi-Strauss, and the Cybernetic Apparatus," *Critical Inquiry* 38, no. 1 (2011): 96–126.

33 François Dosse, *History of Structuralism: The Rising Sign 1945–1966*, translated by Deborah Glassman (Minneapolis: University of Minnesota Press, 1998), n.p.

34 Claude Lévi-Strauss, "Mathematics of Man," *International Social Science Bulletin (UNESCO)* 6, no. 4 (1954): 585.

Chapter 2

1 Reyner Banham, *Theory and Design in the First Machine Age* (London: Architectural Press, 1960), 10, 329–330.

2 H. R. Hitchcock and Philip Johnson, *The International Style* (1932; New York: W. W. Norton, 1995).

3 "Program: The Four Great Makers and the Next Phase in Architecture," 1961, 1. Box 21, Serge Ivan Chermayeff Architectural Records and Papers, 1909–1980, Dept. of Drawings & Archives, Avery Architectural and Fine Arts Library, Columbia University.

4 John Hastings, "Press Release: The Four Great Makers and the Next Phase in Architecture," 1961, n.p. Box 21, Serge Ivan Chermayeff Architectural Records and Papers, 1909–1980, Dept. of Drawings & Archives, Avery Architectural and Fine Arts Library, Columbia University.

5 Ibid.

6 The full title of the cycle in which Chermayeff participated was "The House for the Modern Family: Urban Towers or Suburban Idyls?" and it was chaired by Pratt Institute Department of Architecture dean Olindo Grossi.

7 Serge Chermayeff, "Let Us Not Make Shapes: Let Us Solve Problems," 1961, 263. Box 21, Serge Ivan Chermayeff Architectural Records and Papers, 1909–1980, Dept. of Drawings & Archives, Avery Architectural and Fine Arts Library, Columbia University.

8 Ibid.

9 Ibid., 265, 259

10 Ibid., 263.

11 In his history of the CIAM discourse on urbanism, Eric Mumford provides an excerpt of how CIAM secretary Sigfried Giedion described the goals of CIAM to the Dutch architect and urbanist Cornelis van Eesteren. Chermayeff's urge "let us solve problems" referred to Giedion's fourth goal. The four goals were: "a) to formulate the contemporary program of architecture. b) to advocate the idea of modern architecture. c) to forcefully introduce this idea into technical, economic, and social circles. d) to see to the resolution of architectural problems." Eric Mumford, *The CIAM Discourse on Urbanism, 1928–1960* (Cambridge, MA: MIT Press, 2000), 10.

12 Christopher Alexander, "The Design of an Urban House and Ways of Clustering It, 1959." Box 4, Folder "Alexander, Christopher, 1958–1966," Serge Ivan Chermayeff Architectural Records and Papers, 1909–1980, Dept. of Drawings & Archives, Avery Architectural and Fine Arts Library, Columbia University.

13 Christopher Alexander, "Information and an Organized Process of Design," in *New Building Research Spring 1961* (Washington, DC: National Academy of Sciences—National Research Council, 1962), 117.

14 Lionel March, Peter Dickens, and Marcial Echenique, "Models of Environment: Polemic for a Structural Revolution," *Architectural Design* 71, no. 5 (1971): 275.

15 Ibid.

16 Ibid.

17 Marcial Echenique, interview with Theodora Vardouli, 2016, published as an appendix in Theodora Vardouli, "Graphing Theory: New Mathematics, Design, and the Participatory Turn," dissertation, Massachusetts Institute of Technology, 2017, 454.

18 Leslie Martin, "Architect's Approach to Architecture," *RIBA Journal* 74, no. 5 (1967): 191–200.

19 Ibid., 191.

20 Ibid., 192.

21 Ibid.

22 For a detailed history of the Whitehall redesign see Adam Sharr and Stephen Thornton, *Demolishing Whitehall: Leslie Martin, Harold Wilson and the Architecture of White Heat* (London: Routledge, 2016).

23 Leslie Martin and Lionel March, "Introduction to Part 1: Explorations," in Martin and March, eds., *Urban Space and Structures* (New York: Cambridge University Press, 1972), 1.

24 Leslie Martin, "RIBA Conference on Architectural Education: Report by the Chairman, Sir Leslie Martin," *Architects' Journal* 127 (1958): 773.

25 For historical accounts of the LUBFS Centre see also Sean Keller, *Automatic Architecture: Motivating Form after Modernism* (Chicago: University of Chicago Press, 2017); Sean Keller, "Fenland Tech: Architectural Science in Postwar Cambridge," *Grey Room* (2006): 40–65; Mary Lou Lobsinger, "Two Cambridges: Models, Methods, Systems and Expertise," in *A Second Modernism: MIT, Architecture, and the "Techno-Social" Moment*, edited by Arindam Dutta (Cambridge, MA: MIT Press, 2013), 652–685; Altino João Magalhães Rocha, "Architecture Theory, 1960–1980: Emergence of a Computational Perspective," dissertation, Massachusetts Institute of Technology, 2004.

26 Lionel March, interview with Theodora Vardouli, 2016, published as an appendix in Vardouli, "Graphing Theory," 431.

27 Ibid., 431–432.

28 For a more extensive summary of Leslie Martin's work see his multiauthored obituary in *Architectural Research Quarterly*: Bernard Levin, "Leslie Martin," *Arq: Architectural Research Quarterly* 4, no. 4 (2000): 295–308.

29 Lionel March, "Research and Environmental Studies," *Cambridge Review* 94, no. 2211 (1973): 90.

30 Lionel March, "Modern Movement to Vitruvius: Themes of Education and Research," *Royal Institute of British Architects Journal* 81, no. 3 (1972): 107; March, "Research and Environmental Studies," 91.

31 March, "Research and Environmental Studies," 87; March, "Modern Movement to Vitruvius," 102.

32 See for example, Marcial Echenique, "Urban Systems: Towards an Explorative Model," 1968, LUBFS Working Paper 7; Marcial Echenique, David Crowther, and Walton Lindsay, "A Structural Comparison of Three Generations of New Towns," 1969, LUBFS Working Paper 25; Philip Tabor, "Pedestrian Circulation in Offices," 1969, LUBFS Working Paper 17; Nicholas Bullock, "An Approach to the Simulation of Activities: A University Example," 1970, LUBFS Working Paper 21; Peter Dickens, "The Location of University Facilities: Explorations," 1969, LUBFS Working Paper 22; Philip Steadman, "The Automatic Generation of Minimum-Standard House Plans," 1970, LUBFS Working Paper 23.

33 By 1973, the LUBFS Centre had received nearly £250,000 in external funds for research and consisted of 18 full-time researchers from several disciplines with a similar number of postgraduates, doctoral candidates, and international visiting scholars (March, "Research and Environmental Studies," 90). The inaugural funding came from the Centre for Environmental Studies, an independent charitable trust established by the UK Ministry of Housing and Local Government. The Ford Foundation made a grant of $750,000 to the Centre for Environmental Studies, which was channeled in part to other research initiatives, including the LUBFS Centre. The Centre for Environmental Studies continued to support the Urban Systems Study (1967–1973) led by Marcial Echenique. Other research projects were funded by other external resources, which are listed on page 322 of the special issue "Models of Environment" in *Architectural Design*: the Gulbenkian Foundation and the UK Department of Education and Science funded the Universities Study (1965–1966, 1967–1968, 1968–1971)

by Steadman, Dickens, and Bullock under Leslie Martin's guidance; and the Ministry of Public Building and Works funded the Office Study (1967–1970) and its supporting Computer Aided Design Study (1969–1970), which included Philip Tabor, Dean Hawkes, and others. LUBFS Centre studies also received funding from the Department of Health and Social Security, the Housing Research Foundation, the Ministry of Housing and Local Government, the Royal Institute of British Architects, the University Grants Committee, and Wates Limited.

34 The image of the "architect-planner" was already present in the CIAM meetings, where leading figures such as Le Corbusier, Sigfried Giedion, and Walter Gropius developed principles that linked the design of buildings to the design of entire cities. Their critics and successors, such as members of the architectural group Team 10, developed ways to think about architecture as a trans-scalar endeavor through "scales of association" and the "structure of the human habitat." Yet these efforts did not find their institutional expression in Anglo-American architectural academia until the 1960s with the establishment of "urban design" as a field of study. See Eric Mumford, "The Emergence of Urban Design in the Breakup of CIAM," in *Urban Design*, edited by Alex Krieger and William S. Saunders (Minneapolis: University of Minnesota Press, 2009), 15–37. For a history of the gradual establishment of urban design as a field in the anglophone world through academic, public policy, and critical discourse channels, see Clément Orillard, "The Transnational Building of Urban Design: Interplay between Genres of Discourse in the Anglophone World," *Planning Perspectives* 29, no. 2 (2014): 209–229.

35 March, interview with Vardouli, in Vardouli, "Graphing Theory," 433.

36 Marcial Echenique, "Models: A Discussion," in Martin and March, *Urban Space and Structures*, 164–174. The chapter came out of the eponymous LUBFS Working Paper 6, published in March 1968.

37 Echenique, "Models: A Discussion" (working paper), n.p.

38 Bullock, Dickens, and Steadman, "The Use of Models in Planning and the Architectural Design Process," 107–108.

39 Lionel March, "Elementary Models of Built Forms," in Martin and March, *Urban Space and Structures*, 56.

40 Leslie Martin and Lionel March, "Introduction to Part 2: Activities, Space and Location," in Martin and March, *Urban Space and Structures*, 109–112.

41 Leslie Martin, Lionel March, and others, "Speculations," in Martin and March, *Urban Space and Structures*, 40.

42 March, "Research and Environmental Studies," 89.

43 Philip Steadman, "Research in Architecture and Urban Studies at Cambridge in the 1960s and 1970s: What Really Happened," *Journal of Architecture* 21, no. 2 (2016): 295. Kenneth Boulding is referenced in LUBFS Working Paper 17 (Tabor, "Pedestrian Circulation in Offices").

44 Kenneth Boulding, "General Systems Theory—The Skeleton of Science," *Management Science* 2, no. 3 (1956): 197. On "spaceship earth" see Kenneth Boulding, "The Economics of the Coming Spaceship Earth," in *Environmental Quality in a Growing Economy: Essays from the Sixth RFF Forum*, edited by H. Jarrett (Baltimore: Johns Hopkins University Press, 1966), 3–14.

45 Boulding, "General Systems Theory," 208, 197.

46 March, "Research and Environmental Studies," 89.

47 Sean Keller, for instance, has used the term "systems aesthetics" to refer to what he views as March's efforts to "reground architecture in mathematical rather than visual form." For Keller, "systems aesthetics" denotes the displacement of aesthetics by a "totalizing science of systems" but also "the anesthetization of systems." See Keller, *Automatic Architecture*, 62; Sean Blair Keller, "Systems Aesthetics: Architectural Theory at the University of Cambridge 1960–75," dissertation, Harvard University, 2005.

48 March, "Modern Movement to Vitruvius," 102.

49 Lionel March and Philip Steadman, *The Geometry of Environment: An Introduction to Spatial Organization in Design* (Cambridge, MA: MIT Press, 1974), 337.

50 Ibid., 8.

51 An abridged version of the section on the new maths and the LUBFS Centre also appears in Theodora Vardouli, "Skeletons, Shapes, and the Shift from Surface to Structure in Architectural Geometry," *Nexus Network Journal* 22 (2020): 487–505.
52 March and Steadman, *The Geometry of Environment*, 341.
53 Colin Rowe, "The Mathematics of the Ideal Villa: Palladio and Le Corbusier Compared," *Architectural Review* (March 1947): 101–104; Rudolf Wittkower, *Architectural Principles in the Age of Humanism* (London: Warburg Institute, 1949).
54 Lionel March, "Mathematics and Architecture since 1960," in *Architecture and Mathematics from Antiquity to the Future*, edited by Kim Williams and Michael Ostwald (Basel: Birkhäuser, 2015), 576. The essay was originally published in *Nexus IV / Centro de Matemática e Aplicações Fundamentais, University of Lisbon*, edited by José Francisco Rodrigues and Kim Williams (Fucecchio [Florence]: Kim Williams Books), 9–33.
55 Ibid.
56 Ibid.
57 Leslie Martin, foreword to March and Steadman, *The Geometry of Environment*, 6.
58 March and Steadman, *The Geometry of Environment*, 7.
59 Ibid.
60 Ibid., 8.
61 Ibid., 6.
62 Nuffield Mathematics Project, *Environmental Geometry* (London: Chambers and Murray, 1969), 1.
63 Ibid., 39.
64 Bryan Thwaites, ed., *On Teaching Mathematics: A Report on Some Present-Day Problems in the Teaching of Mathematics Being the Outcome of the Discussions and Lectures at the Southampton Mathematical Conference 1961* (New York: Pergamon Press, 1961), xiv.
65 On the trajectory of the "new math" in the United States, see Christopher J. Phillips, *The New Math: A Political History* (Chicago: University of Chicago Press, 2014). Phillips's account exposes the new math's political entanglements and motivations. He shows that during the Cold War, governmental entities and the public put emphasis on mathematical training with the aim to cultivate in citizens an aptitude for formal and structural reasoning. In that context, traditional mathematical pedagogical practices of memorization and manipulation were seen as cultivating mechanical habits, which countered the ideal of the active and free-minded US citizen (135).
66 Thwaites, *On Teaching Mathematics*, 60.
67 Bryan Thwaites, *The School Mathematics Project: The First Ten Years* (London: Cambridge University Press, 1972), 3.
68 Tim Wheatley, "S.M.P. and Curriculum Development," in Thwaites, *The School Mathematics Project*, 224.
69 Douglas Quadling, "The Mathematics of S.M.P.," in Thwaites, *The School Mathematics Project*, 217.
70 Ibid.
71 Thwaites, *On Teaching Mathematics*, 28.
72 Quadling, "The Mathematics of S.M.P.," 218.
73 Ibid.
74 Organisation for European Economic Co-operation and Office for Scientific and Technical Personnel, "Professor Dieudonné's Address," in *New Thinking in School Mathematics* (Paris, 1961), 35. The original text is "A bas Euclid!," which translates as "Down with Euclid!"
75 Ibid., 41.
76 On ancient Greek geometry as a form of visual practice see José Ferreirós, *Mathematical Knowledge and the Interplay of Practices* (Princeton, NJ: Princeton University Press, 2015). In his account of the US new math, Phillips also emphasizes the break from acts of drawing and interpreting figures in the teaching of geometry and the shift toward systematic reasoning with axioms and theorems (Phillips, *The New Math*, 63).
77 Thwaites, *The School Mathematics Project*, 18.
78 Ibid., 29.
79 See Jeremy Gray on Herbert Mehrtens's untranslated work, which juxtaposed Klein and Hilbert in relationship to modernist ideals in mathematics. J. J. Gray, "Modern Mathematics as a Cultural Phenomenon," in *Architecture of Modern Mathematics: Essays in History and Philosophy*, edited by J. Ferreirós and J. J. Gray (Oxford: Oxford University Press, 2006), 371–396.

80 Klein's 1872 Erlangen lecture was titled "Vergleichende Betrachtungen über neuere geometrische Forschungen" (Comparative reviews of recent researches in geometry).
81 Jeremy Gray, *Plato's Ghost: The Modernist Transformation of Mathematics* (Princeton, NJ: Princeton University Press, 2008), 10.
82 Quadling, "The Mathematics of S.M.P.," 219.
83 Thwaites, *The School Mathematics Project*, 1.
84 Philip Steadman, interview with Theodora Vardouli, 2016, published as an appendix in Vardouli, "Graphing Theory,"443; Echenique, interview with Vardouli, in ibid., 453.
85 Steadman, interview with Vardouli, in Vardouli, "Graphing Theory," 449.
86 March, interview with Vardouli, in Vardouli, "Graphing Theory," 431.
87 Steadman, interview with Vardouli, in Vardouli, "Graphing Theory," 449.
88 W. T. Tutte, "Commentary," in Dénes König, *Theory of Finite and Infinite Graphs*, translated by Richard McCoart (Basel: Birkhäuser, 1990), 1.
89 Ibid.
90 A. Kaufmann, *Points and Arrows: The Theory of Graphs*, translated by Richard Sadler (London: Transworld, 1972), 134.
91 Frank Harary, Robert Z. Norman, and Dorwin Cartwright, *Structural Models: An Introduction to the Theory of Directed Graphs* (New York: John Wiley & Sons, 1965), v.
92 Frank Harary, "On the ABC of Graph Applications," *Le Matematiche* 45, no. 1 (1991): 75.
93 March and Steadman, *The Geometry of Environment*, 13.
94 Ibid., 20.
95 Ibid., 23.
96 Ibid., 27.
97 Ibid., 243.
98 Ibid., 12.
99 Ibid., 257 (emphasis in the original).
100 Ibid., 260.
101 Ibid., 256.
102 Ibid., 262.
103 Ibid. (emphasis in the original).
104 Ibid., 268.
105 A short reference to this material also appears in Theodora Vardouli, "Bioptemes and Mechy Max Systems: Topological Imaginations of Adaptive Architecture," in *Disruptive Technologies: The Convergence of New Paradigms in Architecture*, edited by Philippe Morel and Henriette Bier (Cham, Switzerland: Springer, 2023).
106 Steadman, "The Automatic Generation of Minimum-Standard House Plans."
107 National Building Agency, *Generic Plans, Two and Three Storey Houses* (London: NBA, 1965); National Building Agency, *Metric House Shells, Two Storey and Metric House Shells, Two Storey, Technical Supplement* (London: NBA, 1968).
108 Ministry of Housing and Local Government, *Homes for Today and Tomorrow, Report of the Parker Morris Committee* (London: HMSO, 1961); Ministry of Housing and Local Government, *Space in the Home, Metric Edition* (London: HMSO, 1968).
109 Steadman, "The Automatic Generation of Minimum-Standard House Plans," 21.
110 R. L. Brooks, C. A. B. Smith, A. H. Stone, and W. T. Tutte, "The Dissection of Rectangles into Squares," *Duke Mathematical Journal* 7, no. 1 (1940): 312–340.
111 Steadman, "The Automatic Generation of Minimum-Standard House Plans," 1.
112 Steadman, "Research in Architecture and Urban Studies at Cambridge in the 1960s and 1970s: What Really Happened," 291.
113 Yona Friedman, *Toward a Scientific Architecture*, translated by Cynthia Lang (Cambridge, MA: MIT Press, 1975), xi.
114 On 18 October 1967, Friedman wrote to Japanese Prime Minister Eisaku Satō to advocate against the impending demolition of the Imperial Hotel in Tokyo. Fond Denise et Yona Friedman, Folder 284, p. 227.
115 Yona Friedman, *L'architecture mobile: vers une cité conçue par ses habitants* (Paris: Casterman, 1970).
116 For collections and interpretive accounts of Friedman's work, see for example Helene Fentener van Vlissingen, Sabine Lebesque, and Yona Friedman, *Structures Serving the Unpredictable* (Rotterdam: NAi Publishers, 1999); Yona Friedman and Manuel Orazi, *Yona Friedman: The Dilution of Architecture*, ed. Nader Seraj (Chicago: University of Chicago Press, 2015).
117 On the entanglements of axiomatic thinking with postwar art and architecture see Alma Steingart, "The Axiom of High Modernism," *Representations* 156, no. 1 (2021): 115–142; Alma

Steingart, "The Axiomatic Aesthetic," in *Computer Architectures: Constructing the Common Ground*, edited by Theodora Vardouli and Olga Touloumi (London: Routledge, 2020), 194–213.

118 Friedman, *Toward a Scientific Architecture*, 23.
119 Ibid., 24.
120 Ibid., 26.
121 Ibid., 24.
122 Ibid., 31.
123 Ibid., 35.
124 Ibid., 39.
125 Ibid., 44.
126 Ibid.
127 Ibid., 46.
128 Ibid., 48.
129 Ibid., 45.
130 Ibid.
131 Frank Harary and Anthony Hill, "On the Number of Crossings in a Complete Graph," *Proceedings of the Edinburgh Mathematical Society* 13, no. 4 (1963): 333–338.
132 Anthony Hill, ed., *DATA: Directions in Art, Theory, and Aesthetics: An Anthology* (Greenwich, CT: New York Graphic Society, 1969).
133 Hill, *DATA*, foreword (n.p.). For a detailed discussion of *Circle* and *DATA* within their cultural and intellectual contexts, see Matthew Allen, *Flowcharting: From Abstractionism to Algorithmics in Art and Architecture* (Zurich: gta Verlag, 2023).
134 Hill, foreword to *DATA*.
135 Yona Friedman, "Reflections on the Architecture of the Future: Criteria for Town-Planning," in Hill, *DATA*.
136 Anthony Hill, "Programme. Paragram. Structure," in Hill, *DATA*, 254.
137 On information aesthetics see also "Computer Art and Information Aesthetics: A Conversation with Leslie Mezei and Frieder Nake Moderated by Theodora Vardouli," in *Designing the Computational Image, Imagining Computational Design*, edited by Daniel Cardoso Llach and Theodora Vardouli (San Francisco: Applied Research and Design Publishing, 2023), 218–235.
138 Hill, "Programme. Paragram. Structure," 264.
139 Correspondence between Friedman and Hill, 5, 7, 12, 20 September 1964, Fond Denise et Yona Friedman, Folder 283, pp. 254, 227, 218, 211.
140 Yona Friedman, letter to Anthony Hill, 20 September 1964, Fond Denise et Yona Friedman, Folder 283, 211.
141 Yona Friedman, letter to Anthony Hill, 10 October 1964, Fond Denise et Yona Friedman, Folder 283, p. 172.
142 Anthony Hill, letter to Yona Friedman, 13 October 1964, Fond Denise et Yona Friedman, Folder 283, pp. 161–162.
143 Anthony Hill, letter to Yona Friedman, 29 January 1967, Fond Denise et Yona Friedman, Folder 284, p. 896. Later correspondence with Elizabeth Case, executive editor, University of Michigan Press indicates that the Press chose to pursue the publication (8 November 1967, Fond Denise et Yona Friedman, Folder 284, p. 136).
144 Yona Friedman, letter to Anthony Hill, 1 February 1967, Fond Denise et Yona Friedman, Folder 284, p. 895.
145 Anthony Hill, letter to Yona Friedman, 11 November 1967, Fond Denise et Yona Friedman, Folder 284, p. 142.
146 Yona Friedman, "Seminar on Methods for Architects and Planners," *Arch+* 2 (1968): 27–44.
147 Ibid., 27.
148 Ibid.
149 Ibid., 28.
150 Ibid.
151 Friedman, *Toward a Scientific Architecture*, 11.
152 Ibid., 12.
153 Ibid.
154 Ibid., 14.
155 König, *Theory of Finite and Infinite Graphs*, 47.
156 Frank Harary, "Aesthetic Tree Patterns in Graph Theory," *Leonardo* 4, no. 3 (1971): 227.
157 Philippe Sers, preface to Yona Friedman, *Pour l'architecture scientifique* (Paris: Belfond, 1971), 7.
158 Larry Busbea offers a rich history of the avant-garde movement of "spatial urbanism" in France. He argues that, through seemingly fanciful and fantastical experiments, art and architecture captured and (pre)figured fundamental social and cultural shifts. See Larry Busbea, *Topologies: The Urban Utopia in France, 1960–1970* (Cambridge, MA: MIT Press, 2007).
159 Eckhard Schulze-Fielitz, "Une théorie pour l'occupation de l'espace," *Architecture d'aujourd'hui* 33, no. 102 (1962): 78, 80.

160 Friedman, *Toward a Scientific Architecture*, xii.

161 A version of this argument and part of the material I present here has been published in Theodora Vardouli, "Points and Lines, Nodes and Rods: Megastructure, Graph Realism, and Yona Friedman's Scientific Architecture," *Arq: Architectural Research Quarterly* 25, no. 3 (2021): 245–254.

162 Bernard Huet, "Folkestone 1966," *Melp! (Melpomene)* 22 (1966): n.p.

163 Metropole Arts Centre, *International Exhibition of Experimental Architecture: The New Metropole Arts Centre, Folkestone, 6–30 June 1966* (Folkestone, UK: Metropole Arts Centre, 1966).

164 Huet, "Folkestone 1966."

165 Friedman, *Toward a Scientific Architecture*, xii.

166 Roger Scruton, "The Architecture of Stalinism," *Cambridge Review* 99 (26 November 1976): 36–41.

167 Ibid., 38.

168 Ibid., 41.

169 Ibid., 37, 41.

170 Philip Steadman, William J. Mitchell, and Dean Hawkes, "The Architecture of Stalinism: A Reply to Dr. Scruton," *Cambridge Review* 99, no. 2237 (1977): 106–112.

171 Ibid., 110.

172 Ibid.

173 Ibid., 107.

Chapter 3

1 This interlude and parts of this chapter draw material from Theodora Vardouli and David Theodore, "Walking Instead of Working: Space Allocation, Automatic Architecture, and the Abstraction of Hospital Labor," *IEEE Annals of the History of Computing* 43, no. 2 (2021): 6–17.

2 Karin Knorr Cetina distinguishes "objects of knowledge" from commodities and instruments. "Objects of knowledge," she argues, are open-ended and indefinitely unfolding; their incompleteness "structures desire" and "provides for the continuation of structures of wanting." See Karin Knorr Cetina, "Sociality with Objects: Social Relations in Postsocial Knowledge Societies," *Theory, Culture, and Society* 14, no. 4 (1997): 13.

3 On the disciplinary stakes of the pursuit for an "automatic architecture" see Sean Keller, *Automatic Architecture: Motivating Form after Modernism* (Chicago: University of Chicago Press, 2017). Keller suggests that the quest for systematic methods and algorithmic design procedures was germane to architecture as a modern discipline, constituted through reflexive self-criticality and anxiety over the legitimacy of its actors and products. He reads "automatic architecture" as a deliberate effort to displace architecture from the realm of culture and politics to the domain of objective fact.

4 Nigel Cross, *The Automated Architect* (London: Pion, 1977).

5 Recent publications contemplating meanings, techniques, and potentials of machine learning for architecture include Stanislas Chaillou, *Artificial Intelligence and Architecture: From Research to Practice* (Basel: Birkhäuser, 2022); Philip Bernstein, *Machine Learning: Architecture in the Age of Artificial Intelligence* (London: RIBA Publishing, 2022); Matias del Campo, *Neural Architecture: Design and Artificial Intelligence* (Novato, CA: ORO Editions, 2022); Matias del Campo and Neil Leach, eds., "Machine Hallucinations: Architecture and Artificial Intelligence," *Architectural Design* 92, no. 3 (2022); Neil Leach, *Architecture in the Age of Artificial Intelligence: An Introduction to AI for Architects* (London: Bloomsbury Visual Arts, 2022); Imdat As and Prithwish Basu, eds., *The Routledge Companion to Artificial Intelligence in Architecture* (London: Routledge, 2021). See also Mario Carpo's historical reflections and future projections on agendas of design and automation in relation to scale and artisanship in *Beyond Digital: Design and Automation at the End of Modernity* (Cambridge, MA: MIT Press, 2023).

6 Goda Klumbytė and Loren Britton remind us that algorithmic abstraction is in fact *extraction* that stems from, but also acts to transform, a concrete thing or setting. Goda Klumbytė and Loren Britton, "Abstracting Otherwise: In Search for a Common Strategy for Arts and Computing," *ASAP/Journal* 5, no. 1 (2020): 24. Kathleen Daly Weisse, Julie Jung, and Kellie Sharp-Hoskins write on algorithmic abstraction as a politically oppressive technique that upholds

white suprematist and culturally imperialist notions of universality. See Kathleen Daly Weisse, Julie Jung, and Kellie Sharp-Hoskins, "Algorithmic Abstraction and the Racial Neoliberal Rhetorics of 23andMe," *Rhetoric Review* 40, no. 3 (2021): 284–299.

7 Yanni Loukissas advocates for the irreducible locality of data and the need to shift attention from data sets to data *settings*. See Yanni Loukissas, *All Data Are Local: Thinking Critically in a Data-Driven Society* (Cambridge, MA: MIT Press, 2019).

8 Leslie Martin, "RIBA Conference on Architectural Education: Report by the Chairman, Sir Leslie Martin," *Architects' Journal* 127 (1958): 773.

9 Ibid.

10 George Kasaboff, John Outram, Paul Power, and Ian McKechnie, "Student Section—BASA," *Architects' Journal* 129 (1959): 451.

11 Christopher Alexander and D. V. Doshi, "A Role for the Individual in City Planning: Main Structure Concept," *Landscape* 13, no. 2 (1963): 17–18.

12 On the tropes and agendas of postwar architectural "research" see Avigail Sachs, "The Postwar Legacy of Architectural Research," *Journal of Architectural Education* 62, no. 3 (2009): 55.

13 Christopher Alexander, "The Revolution Finished Twenty Years Ago," *Architect's Year Book* 9 (1960): 181.

14 Roger Montgomery, "Pattern Language—The Contribution of Christopher Alexander's Centre for Environmental Structure to the Science of Design," *Architectural Forum* 132, no. 1 (1970): 52.

15 Christopher Alexander, *Notes on the Synthesis of Form* (Cambridge, MA: Harvard University Press, 1964).

16 For histories of design methods and design research see Nigan Bayazit, "Investigating Design: A Review of Forty Years of Design Research," *Design Issues* 20, no. 1 (2004): 16–29; Geoffrey Broadbent, *Design in Architecture: Architecture and the Human Sciences* (1973; London: Fulton, 1988); Nigel Cross, "Forty Years of Design Research," *Design Research Quarterly* 1, no. 2 (2006): 3–5; Victor Margolin, "Design Research: Towards a History," in *DRS International Conference "Design and Complexity,"* edited by P. Lalande (Montreal: University of Montreal, 2010); Alise Upitis, "Nature Normative: The Design Methods Movement, 1944–1967," dissertation, Massachusetts Institute of Technology, 2008.

17 Allan B. Reswick, foreword to *Proceedings of the First Conference on Engineering Design Education* (Cleveland: Office of Special Programs, Case Institute of Technology, 1960), i–vi.

18 Morris Asimow, *Introduction to Design* (Englewood Cliffs, NJ: Prentice-Hall, 1962); Fritz Zwicky, *Discovery, Invention, Research through the Morphological Approach* (Toronto: Macmillan, 1969).

19 Broadbent, *Design in Architecture*, 272.

20 Ibid.

21 R. S. Easterby, "Review of 'Conference on Design Methods,'" *Ergonomics* 7, no. 4 (1964): 500; William R. Spillers, foreword to Spillers, ed., *Basic Questions of Design Theory* (Amsterdam: North-Holland, 1974).

22 Nathan Ensmenger reads flowcharts—the schematic diagrams of a computer program's logical structure—through Kathryn Henderson's concept of "conscription devices": boundary objects that enlist participation by multiple parties by serving as repositories of knowledge and interaction. See Nathan Ensmenger, "The Multiple Meanings of a Flowchart," *Information and Culture: A Journal of History* 51, no. 3 (2016): 321–351; Kathryn Henderson, "Flexible Sketches and Inflexible Data Bases: Visual Communication, Conscription Devices, and Boundary Objects in Design Engineering," *Science, Technology, and Human Values* 16, no. 4 (1991): 448–473.

23 Cross, *Automated Architect*, 24.

24 Alexander, *Notes on the Synthesis of Form*, 4. See also Theodora Vardouli, "'Bewildered, the Form-Maker Stands Alone': Computer Architecture and the Quest for Design Rationality," in *Computer Architectures: Constructing the Common Ground*, edited by Theodora Vardouli and Olga Touloumi (London: Routledge, 2020), 58–76.

25 "Computers in Building: PERT and CMP," *Architects' Journal Information Library* (1962): 1329.

26 Ibid., 1330 (emphasis in the original).

27 Ibid.

28 "Computers in Action: Ferranti Demonstrates Transportation Technique," *Architects' Journal* 137, no. 18 (1963): 928.

29 Bruce Archer, "Letters: Computers in Building," *Architects' Journal* 138, no. 5 (1963): 214.

30 Stephen Boyd Davis and Simone Gristwood, "'A Dialogue between the Real-World and the Operational Model': The Realities of Design in Bruce Archer's 1968 Doctoral Thesis," *Design Studies* 56 (2018): 185–204.

31 Archer, "Letters: Computers in Building."

32 "Computers in Building: Planning Accommodation for Hospitals and the Transportation Problem Technique," *Architects' Journal Information Library* (1963): 139–142.

33 See also Alistair Fair, "'Modernization of Our Hospital System': The National Health Service, the Hospital Plan, and the 'Harness' Programme, 1962–77," *Twentieth Century British History* 29, no. 4 (2018): 547–575.

34 "Computers in Building: Planning Accommodation for Hospitals and the Transportation Problem Technique," 139. On the metaphor of the "slave" in computer-aided design see Daniel Cardoso Llach, "Perfect Slaves and Cooperative Partners: Steven A. Coons and Computers' New Role in Design," in Cardoso Llach, *Builders of the Vision: Software and the Imagination of Design* (New York: Routledge, 2015).

35 "Computers in Building: Planning Accommodation for Hospitals and the Transportation Problem Technique," 141.

36 Ibid., 142.

37 Ibid.

38 B. Whitehead, "Letters: Computers in Building," *Architects' Journal* 138, no. 5 (1963): 214.

39 Ibid.

40 "Letters: Computers in Building," 214.

41 B. Whitehead and M. Z. Eldars, "An Approach to the Optimum Layout of Single-Storey Buildings," *Architects' Journal* 139, no. 25 (1964): 1373–1380.

42 David Theodore, "'The Fattest Possible Nurse': Architecture, Computers, and Postwar Nursing," in *Daily Life in Hospital: Theories and Practices from the Medieval to the Modern*, edited by Laurinda Abreu and Sally Sheard (Oxford: Peter Lang, 2013), 273–298.

43 *Studies in the Functions and Design of Hospitals* (London: Nuffield Trust, 1956).

44 Richard Muther, *Systematic Layout Planning* (Boston: Cahners Books, 1961).

45 Jon Agar has argued that the algorithmic imagination and systematization of certain practices is a precondition of their computer automation. See Jon Agar, "What Difference Did Computers Make?," *Social Studies of Science* 36, no. 6 (2006): 869–907.

46 One key development in computer-aided planning in the United States was Coplanner. This was a computer program implemented in the context of a project on Collaborative Research in Hospital Planning of the United States Public Health Service and published by the American Hospital Association. This computer program, designed in collaboration J. C. R. Licklider, former vice president of the computer consultancy Bolt Beranek and Newman (BBN) and newly appointed head of the Information Processing Techniques Office (IPTO) at the Defense Department's Advanced Research Projects Agency (ARPA), used similar methods of calculating the "costs" of hospital layouts proposed by the planner to evaluate and rank them—not to algorithmically derive optimum layouts. This reflected research priorities of time-sharing and interactive computing spearheaded by Licklider, as well as input and output devices such as oscilloscopes and light pens supported of the Digital Equipment Corporation PDP-1 computer that ran the program. In their influential article "Computer Technology: A New Tool for Planning," BBN members J. J. Souder and W. E. Clark discussed Coplanner as the future direction of computer-aided planning. J. J. Souder and W. E. Clark, "Computer Technology: A New Tool for Planning," *AIA Journal* (1963): 99.

47 A program for automatic layout planning that developed in the United States around the same time as the Whitehead and Eldars program was CRAFT. This also focused on layout optimization, but, unlike Whitehead and Eldars's which followed an additive method of placing one activity after another, CRAFT used what was described by architectural researchers

of the time as a "permutational" method: it began with an arbitrary room layout and then exchanged pairs of rooms until an exchange that optimized distance or cost had been found. This exchange was then "fixed" and other pairs exchanged until the system reached equilibrium. See E. S. Buffa, G. S. Armour, and T. E. Vollman, "Allocating Facilities with CRAFT," *Harvard Business Review* 42, no. 2 (1963): 136–140.

48 See, for example, T. Willoughby, W. Paterson, and G. Drummond, "Computer-Aided Architectural Planning," *Operational Research Quarterly* 21, no. 1 (1970): 91–98; W. J. Mitchell, Philip Steadman, and Robin S. Liggett, "Synthesis and Optimization of Small Rectangular Floor Plans," *Environment and Planning B: Planning and Design* 3, no. 1 (1976): 37–70; William J. Mitchell, "The Automated Generation of Architectural Form," in *Proceedings of the 8th Design Automation Workshop* (New York: ACM, 1971), 193–207.

49 T. W. Maver, "The Computer as an Aid to Architectural Design: Present and Future," 1970, ABACUS Occasional Paper 10, University of Strathclyde, 15.

50 Ministry of Public Building and Works Committee on the Application of Computers in the Construction Industry, *Computers for Contractors in the Building Industry: First Report of the Sub-Committee on Computer Uses in the Construction of Buildings to the Committee on the Application of Computers in the Construction Industry* (London: MPBW, Directorate of Research and Information, 1969); Donald Gibson, ed., *Construction, Education and the Computer: Conference Report* (London: MPBW, Directorate of Research and Information, 1969); Annette Stannett and Ministry of Public Building and Works Committee on the Application of Computers in the Construction Industry, *Bibliography on the Application of Computers in the Construction Industry 1962–1967* (London: HMSO, 1968).

51 "Models of Environment," *Architectural Design* 71, no. 5 (1971): 322.

52 Philip Tabor, "Pedestrian Circulation in Offices," 1969, 60, LUBFS Working Paper 17.

53 Ibid.

54 Lionel March, Peter Dickens, and Marcial Echenique, "Models of Environment: Polemic for a Structural Revolution," *Architectural Design* 71, no. 5 (1971): 275.

55 Tabor, "Pedestrian Circulation in Offices"; Philip Tabor, "Traffic in Buildings 2: Systematic Activity Location," 1970, LUBFS Working Paper 18; Philip Tabor, "Traffic in Buildings 3: Analysis of Communication Patterns," 1970, LUBFS Working Paper 19; Philip Tabor, "Traffic in Buildings 4: Evaluation of Routes," 1970, LUBFS Working Paper 21.

56 Tabor, "Pedestrian Circulation in Offices," 19.

57 Ibid., 10.

58 Ibid., 61.

59 Ibid., 51.

60 Tabor, "Traffic in Buildings 2: Systematic Activity Location," 2.32.

61 P. H. Levin, "The Use of Graphs to Decide the Optimum Layout of Buildings," *Architect's Journal* 140 (1964): 809.

62 Ibid., 815.

63 B. Whitehead and M. Z. Eldars, "Optimum Layouts," *Architects' Journal* 140 (1964): 1223.

64 Christopher Alexander, "A Much Asked Question about Computers in Design," 1. Box 4, Folder "Alexander, Christopher, 1958–1966," Serge Ivan Chermayeff Architectural Records and Papers, 1909–1980, Dept. of Drawings & Archives, Avery Architectural and Fine Arts Library, Columbia University.

65 Christopher Alexander, "A Much Asked Question about Computers in Design," *Landscape* (1967): 8–12.

66 Alexander, "A Much Asked Question about Computers in Design" (archival draft), 4, 6.

67 Ibid., 5.

68 Ibid.

69 Ibid., 4.

70 Serge Chermayeff and Christopher Alexander, *Community and Privacy: Toward a New Architecture of Humanism* (Garden City, NY: Doubleday and Company, 1963), 152.

71 Christopher Alexander, "The Design of an Urban House and Ways of Clustering It, 1959," 6. Box 4, Folder "Alexander, Christopher, 1958–1966," Serge Ivan Chermayeff Architectural Records and Papers, 1909–1980, Dept. of Drawings & Archives, Avery Architectural and Fine Arts Library, Columbia University.

72 Christopher Alexander, "Letter to Chermayeff Re: Failures Interlock, IBM Group Extraction," 1960. Box 4, Folder "Alexander, Christopher, 1958–1966," Serge Ivan Chermayeff Architectural Records and Papers, 1909–1980, Dept. of Drawings & Archives, Avery Architectural and Fine Arts Library, Columbia University.
73 Christopher Alexander, "Programme (The Urban House)," 1959. Box 4, Folder "Alexander, Christopher, 1958–1966," Serge Ivan Chermayeff Architectural Records and Papers, 1909–1980, Dept. of Drawings & Archives, Avery Architectural and Fine Arts Library, Columbia University.
74 Christopher Alexander and Marvin L. Manheim, "HIDECS 2: A Computer Program for the Hierarchial Decomposition of a Set Which Has an Associated Linear Graph," 1962, 42, MIT, Civil Engineering Systems Laboratory Publication 160.
75 On a material history of HIDECS 2 see Alise Upitis, "Alexander's Choice: How Architecture Avoided Computer-Aided Design c. 1962," in *A Second Modernism: MIT, Architecture, and the "Techno-Social" Moment*, edited by Arindam Dutta (Cambridge, MA: MIT Press, 2013), 474–506.
76 Alexander, *Notes on the Synthesis of Form*, 26.
77 Alexander, "Failures Interlock."
78 Alexander, "Information and an Organized Process of Design," 117.
79 Alexander, "Programme (The Urban House)."
80 Christopher Alexander, "A City Is Not a Tree," *Architectural Forum* 122 (1965): 58–62.
81 Christopher Alexander, "Letter to Chermayeff Re: Yale Position Offer," 1965. Box 4, Folder "Alexander, Christopher, 1958–1966," Serge Ivan Chermayeff Architectural Records and Papers, 1909–1980, Dept. of Drawings & Archives, Avery Architectural and Fine Arts Library, Columbia University.
82 Ibid.
83 Sara Ishikawa, interview with Theodora Vardouli, 13 October 2016, published as an appendix in Theodora Vardouli, "Graphing Theory: New Mathematics, Design, and the Participatory Turn," dissertation, Massachusetts Institute of Technology, 2017.
84 Serge Chermayeff, "Recommendation Letter for Christopher Alexander to Charles W. Moore," 1963. Box 4, Folder "Christopher Alexander," Serge Ivan Chermayeff Architectural Records and Papers, 1909–1980, Dept. of Drawings & Archives, Avery Architectural and Fine Arts Library, Columbia University.
85 Christopher Alexander, "Draft Sent to Serge Chermayeff," n.d. Box 32, Folder "Ten Year Program for Research on Environmental Design," Serge Ivan Chermayeff Architectural Records and Papers, 1909–1980, Dept. of Drawings & Archives, Avery Architectural and Fine Arts Library, Columbia University.
86 Ibid., 10.
87 Ibid., n.p.
88 Ibid.
89 Ibid., 7.
90 Ibid., n.p.
91 Ibid.
92 Ibid.
93 Ibid.
94 Christopher Alexander, Van Maren King, and Sara Ishikawa, *390 Requirements for the Rapid Transit Station* (Berkeley, CA: Center for Environmental Structure, 1965).
95 Sara Ishikawa, "Origins of Patterns & A Pattern Language: Impressions of Experience," in *Current Challenges for Patterns, Pattern Languages and Sustainability*, edited by Hajo Neis and Gabriel Brown (Portland, OR: PUARL Press, 2009), 17–21.
96 Ibid., 17.
97 Stephen Grabow, *Christopher Alexander: The Search for a New Paradigm in Architecture* (Stockfield, UK: Oriel Press, 1983), 31.
98 Christopher Alexander, "The Coordination of the Urban Rule System," in *Regio Basiliensis Proceedings* (1965), 1–9.
99 Ian Moore and Barry Poyner, Activity Data Method (London: Ministry of Public Building and Works, 1965).
100 Broadbent, *Design in Architecture*, 288.
101 Christopher Alexander and Barry Poyner, *The Atoms of Environmental Structure* (Berkeley: Center for Planning and Development Research, University of California, 1966).
102 Ibid., 2.
103 Ibid., 4.
104 For a critique of behaviorist tenets in the relational theory, see Janet Daley, "A

Philosophical Critique of Behaviourism in Architectural Design," in *Design Methods in Architecture*, edited by Geoffrey Broadbent and Anthony Ward (London: Lund Humphries, 1969), 71–75.
105 Alexander and Poyner, *The Atoms of Environmental Structure*, 2.
106 Ishikawa, "Origins of Patterns," 17; Ishikawa's research on entrance relations of suburban houses was published as "Twenty-Six Entrance Relations for a Suburban House," in *The Atoms of Environmental Structure* (London: Ministry of Public Buildings and Works, Directorate of Research and Development, 1966), 17–73.
107 Christopher Alexander, Van Maren King, Sara Ishikawa, Michael Baker, and Patrick Hyslop, "Relational Complexes in Architecture," *Architectural Record* 140 (1966): 187.
108 Ibid., 186.
109 Ibid, 185.
110 Ibid., 188.
111 Ibid., 190.
112 Gerald Davis, letter to Christopher Alexander, 18 March 1965. Christopher Alexander / Center for Environmental Structure Archives. 1965. VIII.H.1_1.
113 Ibid.
114 Ibid.
115 Ibid.
116 Christopher Alexander, letter to Gerald Davis, 22 March 1965. Christopher Alexander / Center for Environmental Structure Archives. 1965. VIII.H.1_2.
117 Ibid.
118 Ibid.
119 Ibid.
120 Gerald Davis, letter to Christopher Alexander, 22 April 1965. Christopher Alexander / Center for Environmental Structure Archives. 1965. VIII.H.1_4.
121 Gerald Davis, letter to Christopher Alexander, 8 May 1965. Christopher Alexander / Center for Environmental Structure Archives. 1965. VIII.H.1_9.
122 Davis, letter to Alexander, 22 April 1965.
123 Christopher Alexander, letter to Gerald Davis, 28 April 1965. Christopher Alexander / Center for Environmental Structure Archives. 1965. VIII.H.1_2.
124 Davis, letter to Alexander, 8 May 1965.
125 Davis, letter to Alexander, 22 April 1965.
126 Alexander, letter to Davis, 28 April 1965.
127 Gerald Davis, Notes for Discussion, 12 October 1965, GD and CA. Christopher Alexander / Center for Environmental Structure Archives. 1965.VIII.H.1_2.
128 Ishikawa, "Origins of Patterns," 17.
129 Ibid.
130 Montgomery, "Pattern Language," 52.
131 Christopher Alexander, "A City Is Not a Tree," *Architectural Forum* 122 (1965): 58–62.
132 Christopher Alexander, "HIDECS 3: Four Computer Programs for the Hierarchical Decomposition of Systems Which Have an Associated Linear Graph," 1963, Civil Engineering Systems Laboratory, Cambridge, MA, Publication Report No. R63-27.
133 Montgomery, "Pattern Language," 54.
134 Ibid.
135 Christopher Alexander, Sara Ishikawa, and Murray Silverstein, *A Pattern Language Which Generates Multi-Service Centers* (Berkeley, CA: Center for Environmental Structure, 1968). Parts of the same work also appeared in the report: Christopher Alexander, Murray Silverstein, and Sara Ishikawa, "A Sublanguage of 70 Patterns for Multi-Service Centers," 1968, Hunts Point Neighborhood Corporation, Hunts Point, Bronx, NY.
136 Alexander, Ishikawa, and Silverstein, *A Pattern Language Which Generates Multi-Service Centers*, 51.
137 Ibid., 5.
138 Murray Silverstein, "O Rose Thou Art Sick: Reflections on a Pattern Language," in Neis and Brown, *Current Challenges for Patterns, Pattern Languages and Sustainability*, 21.
139 Grabow, *Christopher Alexander*, 55.
140 Silverstein, "O Rose Thou Art Sick," 19.
141 Alexander and Poyner, *The Atoms of Environmental Structure*, 9.
142 Ibid., 17.
143 Grabow, *Christopher Alexander*, 94.
144 Christopher Alexander, "Systems Generating Systems," *Architectural Design* 38 (1968): 605–608.
145 Ibid., 605.
146 Ibid.
147 Montgomery, "Pattern Language," 56.
148 Grabow, *Christopher Alexander*, 54.

149 Michael Lynch, "Pictures of Nothing? Visual Construals in Social Theory," *Sociological Theory* 9, no. 1 (1991): 1–21.
150 Broadbent, *Design in Architecture*, 287.
151 Frank Harary and J. R. Rockey, "A City Is Not a Semilattice Either," *Environment and Planning A* 8 (1976): 375–384.
152 Ibid., 375.
153 bid., 379.
154 Matthew Hunter, "Modeling: A Secret History of Following," in *Design Technics: Archaeologies of Architectural Practice*, edited by Zeynep Çelik Alexander and John May (Minneapolis: University of Minnesota Press, 2019), 45–70.
155 On the circulation and immutability of objects in contexts of knowledge production see Bruno Latour, "Visualisation and Cognition: Drawing Things Together," *Avant: Trends in Interdisciplinary Studies* 3 (2012): 207–260.

Chapter 4

1 Stewart Brand, ed., *The Next Whole Earth Catalog* (Sausalito, CA: Point, 1980).
2 Ernest Callenbach, "A Pattern Language," in Brand, *The Next Whole Earth Catalog*, 217.
3 Ibid.
4 Stewart Brand, "A Pattern Language," in Brand, *The Next Whole Earth Catalog*, 217.
5 On the theoretical and methodological pluralism, or according to some authors confusion, of the concept "participation" see, for example, Sherry Arnstein, "A Ladder of Citizen Participation," *Journal of the American Planning Association* 35, no. 4 (1969): 216–224; Reyner Banham, "Alternative Networks for the Alternative Culture?," in *Design Participation: Proceedings of the Design Research Society's Conference, Manchester, September 1971*, edited by Nigel Cross (London: Academy Editions, 1972), 16; J. Johnson, "A Plain Man's Guide to Participation," *Design Studies* 1, no. 1 (1979): 27–30; Fredrik Wulz, "The Concept of Participation," *Design Studies* 7, no. 1 (1986): 153–162.
6 See Horst Rittel's famous article with Melvin Webber proposing the concept of "wicked problems" in juxtaposition to informationally tractable "tame problems" (Horst Rittel and Melvin Webber, "Dilemmas in a General Theory of Planning," *Policy Sciences* 4 [1973]: 155–169). Rittel and Webber argued for the replacement of problem-solving with problem-setting in the context of a deliberative process. Part of this work was based on Rittel's research on "issues" as key units of an argumentative process and a system for mapping them, called IBIS. See Horst Rittel and Werner Kunz, "Issues as Elements of Information Systems," 1970, Institute of Urban and Regional Development, University of California, Berkeley.
7 Sara Ishikawa, "Origins of Patterns & A Pattern Language: Impressions of Experience," in *Current Challenges for Patterns, Pattern Languages and Sustainability*, edited by Hajo Neis and Gabriel Brown (Portland, OR: PUARL Press, 2009), 18.
8 Christopher Alexander and Center for Environmental Structure, *Houses Generated by Patterns* (Berkeley, CA: Center for Environmental Structure, 1969).
9 Ishikawa, "Origins of Patterns," 19.
10 Christopher Alexander, Shlomo Angel, Sara Ishikawa, Dennis Abrams, and Murray Silverstein, *The Oregon Experiment* (New York: Oxford University Press, 1975). Other major projects that fed into the development of patterns included an 80-panel exhibit for the Japanese Pavilion in the 1970 Osaka world's fair titled "A Human City"; the programming and design for Berkeley City Hall Complex in 1970; and a 20,000-square-foot community mental health center in Modesto, California. The Osaka world's fair exhibition, redesigned by Ron Walkey, consolidated work on urban and community patterns developed by members and students of the Center for Environmental Structure. The Osaka Expo material eventually became the "Towns" section of *A Pattern Language*. See Ishikawa, "Origins of Patterns," 18.
11 Stephen Grabow, *Christopher Alexander: The Search for a New Paradigm in Architecture* (Stockfield, UK: Oriel Press, 1983), 95.
12 Christopher Alexander, *The Process of Creating Life*, vol. 2, *The Nature of Order: An Essay on the Art of Building and the Nature of the Universe* (Berkeley, CA: Center for Environmental Structure, 2001), 303.

13 Grabow, *Christopher Alexander*, 95.
14 Ibid.
15 Ibid.
16 Christopher Alexander and D. V. Doshi, "A Role for the Individual in City Planning: Main Structure Concept," *Landscape* 13, no. 2 (1963): 17.
17 The International Design Conference in Aspen (IDCA), launched in 1951 by Walter Paepcke, aimed to bring together art, design, manufacturing, and business. It became one of the foremost design gatherings of the second half of the twentieth century, and over the years hosted a roster of internationally renowned architects, designers, theorists, and historians. On the IDCA see Reyner Banham and International Design Conference in Aspen, *The Aspen Papers: Twenty Years of Design Theory from the International Design Conference in Aspen* (New York: Praeger, 1974); International Design Conference in Aspen Records, Online Archive of California, https://oac.cdlib.org/findaid/ark:/13030/c8pg1t6j/admin/
18 Alexander and Doshi, "A Role for the Individual in City Planning," 18.
19 Ibid., 20.
20 Ibid., 18.
21 Ibid., 20.
22 Alison Smithson and Team 10, *Team 10 Meetings: 1953–1984* (New York: Rizzoli, 1991).
23 Alise Upitis, "Nature Normative: The Design Methods Movement, 1944–1967," dissertation, Massachusetts Institute of Technology, 2008, 178.
24 N. John Habraken, *Supports: An Alternative to Mass Housing* (1972; Newcastle upon Tyne: Urban International Press, 1999), 74, 94.
25 Guy Weinzapfel and Nicholas Negroponte, "Architecture-by-Yourself: An Experiment with Computer Graphics for House Design," in *Proceedings of the 3rd Annual Conference on Computer Graphics and Interactive Techniques*, SIGGRAPH '76 (New York: ACM, 1976), 74.
26 Ibid.
27 Yona Friedman, "The Flatwriter," in *Pro Domo* (Barcelona: Actar, 2006), 129; reprint of Friedman, "The Flatwriter: Choice by Computer," *Progressive Architecture* 52 (1971): 98–101.
28 Friedman, "The Flatwriter," 130.
29 Ibid.
30 Yona Friedman, *Toward a Scientific Architecture*, translated by Cynthia Lang (Cambridge, MA: MIT Press, 1975), 33.
31 Ibid., 1.
32 Yona Friedman, "A Research Program for a Scientific Method of Planning," *Architectural Design* 37, no. 7 (1967): 380.
33 Yona Friedman, "Towards a Coherent System of Planning," *Architectural Design* 34, no. 8 (1964): 371.
34 Friedman, "A Research Program for a Scientific Method of Planning," 379.
35 Ibid.
36 Ibid, 380.
37 Ibid.
38 Ibid.
39 Ibid., 381.
40 Ibid.
41 Friedman, *Toward a Scientific Architecture*, 54.
42 For a comprehensive discussion of intellectual, technical, and political developments around the notion of the "user" in the context of governmentally funded housing projects in postwar France see Kenny Cupers, *The Social Project: Housing in Postwar France* (Minneapolis: University of Minnesota Press, 2014).
43 See "Programming the Villes Nouvelles," in Cupers, *The Social Project*.
44 Cupers, *The Social Project*, 284.
45 Ibid., 288.
46 Friedman, *Toward a Scientific Architecture*, 35.
47 The title translates as "From Evolutionary Dwelling to Self-Planning." Michel Ragon, "Yona Friedman: De l'habitat évolutif à l'autoplanification," *Urbanisme* 43, no. 143 (1974): 75–77.
48 Ibid., 75.
49 Geoffrey Broadbent and Anthony Ward, *Design Methods in Architecture* (London: Lund Humphries, 1969).
50 Collin Cave and Keith Elvin, "Design Methods: Not Only How but Why," *Architects' Journal* 147 (1968): 63.
51 Cross, *Design Participation*.
52 On the Design Participation conference see also Theodora Vardouli, "Who Designs?," in *Empowering Users through Design*, edited by David Bihanic (Cham, Switzerland: Springer, 2015), 13–41.

53 Yona Friedman, "Information Processes for Participatory Design," in Cross, *Design Participation*, 45–50.

54 Nicholas Negroponte, "Research in Progress: The Architecture Machine," *Computer Aided Design* 7, no. 3 (1975): 190.

55 Nicholas Negroponte, "Aspects of Living in an Architecture Machine," in Cross, *Design Participation*, 63–67.

56 The text on SEEK also appears in part in Theodora Vardouli, "SEEK," in *The Architecture of Closed Worlds, or, What Is the Power of Shit?* by Lydia Kallipoliti (New York: Lars Müller/ Storefront Editions, 2018).

57 Leon Bennett Groisser and Nicholas Negroponte, *Computer Aids to Participatory Architecture* (Cambridge, MA: MIT, 1971), 140.

58 Nicholas Negroponte, *The Architecture Machine: Toward a More Human Environment* Cambridge, MA: MIT Press, 1973), 71.

59 In URBAN5, the architect produced configurations of urban form by essentially inputting graphs: moving around ten-by-ten-foot cubes in an IBM 2250 graphic display with a light pen and typing material and functional assignments in natural language. Despite having volume, the cubes were essentially discrete points corresponding to the end of the tip of the light pen. The computations proceeded based on labels assigned to these points and computations based on their relations. URBAN5 featured six SYMbol buttons, each holding 16 symbols that described activities, functions, or formal qualifications. The user populated each SYM button with existing symbols or defined new ones, by typing a word and relating it to one of 16 predefined generic categories (daily, commercial, education, service, private, etcetera). URBAN5 could accommodate up to 64 symbols assigned to cube units, with each cube accepting only one symbol assignment—every point of the graph essentially accepting one label. See Nicholas Negroponte, "URBAN 5, an On-Line Urban Design Partner," *Ekistics* (1967): 290.

60 Gary Moore, ed., *Emerging Methods in Environmental Design and Planning* (Cambridge, MA: MIT Press, 1973).

61 Nicholas Negroponte and Leon Bennett Groisser, "URBAN5: A Machine That Discusses Urban Design," in Moore, *Emerging Methods in Environmental Design and Planning*, 114.

62 Ibid.

63 Warren M. Brodey, "The Design of Intelligent Environments: Soft Architecture," *Landscape*, no. 17 (1967): 8–12.

64 Friedman, *Toward a Scientific Architecture*, ix.

65 Ibid., ix.

66 Nicholas Negroponte, *Soft Architecture Machines* (Cambridge, MA: MIT Press, 1975), 100.

67 Weinzapfel and Negroponte, "Architecture-by-Yourself," 74. Friedman also wrote on the Architecture-by-Yourself theory and methods in Judith R. Blau, Mark La Gory, and John Pipkin, eds., *Professionals and Urban Form* (Albany: State University of New York Press, 1983).

68 Weinzapfel and Negroponte, "Architecture-by-Yourself," 74.

69 Ibid.

70 Ibid.

71 While in the first half of the 1960s most computer-aided research at MIT transpired under the umbrella of the Computer-Aided Design Project (CAD Project)—a project mainly affiliated with the Departments of Mechanical and Electrical Engineering—the end of the decade found the Department of Architecture establishing its presence in computer aids development. In December 1968 School of Architecture and Planning dean Lawrence Anderson was reporting that despite the comparatively slow adoption of "the powerful tools of computation" in architecture and planning, "the trend is now in full swing." The start of the 1970s found the Department of Architecture harnessing a tenfold increase in computing power and storage capacity of "in-house" computing facilities within a year and a dramatic rise in research funding, from a cash flow of $256 per year in 1965 to $198,255 in 1970. Research activity in architecture and planning at MIT was also boosted by the School's participation in new institute-wide interdisciplinary initiatives such as the Ford Foundation-funded Urban Systems Laboratory (USL), founded in 1968 with the goal to connect, support, and amplify urban research efforts by faculty members and students via computer, information, and other resources.

For a detailed history of the Computer Aided Design Project and its intellectual and technical infrastructures, see Daniel Cardoso Llach, *Builders of the Vision: Software and the Imagination of Design* (New York: Routledge, 2015). This note draws from two reports: Lawrence B. Anderson, "School of Architecture and Planning: Report of the Dean," *Massachusetts Institute of Technology Bulletin* 104, no. 3 (1968): 29–33; and Charles L. Miller, "Urban Systems Laboratory," *Massachusetts Institute of Technology Bulletin* 104, no. 3 (1968): 490.

72 Johnson had previously developed a three-dimensional extension of SKETCHPAD—the first graphical computer-aided design program that came out of Ivan Sutherland's dissertation. Guy Weinzapfel, Timothy E. Johnson, and John Perkins, "IMAGE: An Interactive Computer System for Multi-Constrained Spatial Synthesis," in *Proceedings of the 8th Design Automation Workshop*, DAC '71 (New York: ACM, 1971), 108.

73 William Lyman Porter, Katherine J. Lloyd, and Aaron Fleisher, *Discourse: The Development of a Language and System for Computer Assisted City Design* (Cambridge, MA: MIT, Laboratory for Environmental Studies, Dept. of City and Regional Planning, 1968).

74 Guy Edward Weinzapfel, "The Function of Testing during Architectural Design," thesis, Massachusetts Institute of Technology, 1971. Having pursued his graduate thesis on theoretical formulations of "testing" in architectural design and its implementations in the IMAGE system, Weinzapfel was versed in the possibilities and discontents of automatic constraint-based space planning. He presented a critique of IMAGE in the second International Conference of the Design Research Society that took place in London in August 1973, themed "The Design Activity." Weinzapfel argued that although IMAGE was a definite advancement in computer-aided design, it compromised the designer's success by its rigid syntax and the entrapment in quantitative tasks. He attributed this to a captivating curiosity around computers' design capabilities, which was removed from the needs and concerns of practicing designers. Such phenomena led Weinzapfel to ask a troubling question for computer-aided design, also featured as his presentation's title: see Guy Weinzapfel, "It Might Work, but Will It Help?," in *Proceedings of the Design Activity International Conference* (London, 1973), 1–16.

75 Weinzapfel and Negroponte, "Architecture-by-Yourself," 75.

76 Ibid. The "dual" representation of floor plans had garnered attention through the work of John Grason, an electrical engineer from Carnegie-Mellon University who wrote his dissertation under political scientist Herbert Simon—famous, alongside his contributions to economics that earned him a Nobel Prize in 1978, as an instigator of the field of "artificial intelligence" and an influential decision-making theorist in design. Grason had proposed to represent "space-filling location problems," that is, problems pertaining to the placement of subspaces in a larger space, as graphs. He presented this work in the first conference of the interdisciplinary Design Methods Group at MIT as a superior alternative to grid mapping representations that were used in floor planning. Supporting this primacy was the argument that graphs were both *requirement* diagrams and *form* diagrams—an attribute he deemed essential for any useful design tool. See John Grason, "A Dual Linear Graph Representation for Space-Filling Location Problems of the Floor Plan Type," in Moore, *Emerging Methods in Environmental Design and Planning*, 170–178.

77 The Architecture Machine acquired a touch-sensitive digitizer from the Canadian company Instronics Ltd. in April 1976, on which they performed "fingerpainting" experiments. This work was conducted by Richard Bolt under ARPA funding (contract number MDA-903-76-C-0261) from 1 April 1976 to 30 September 1976. Work on touch-sensitive displays continued in 1977 under Army Research Institute funding (DAHCI9-77-G-0014) with Negroponte as principal investigator. For an overview, see Christopher F. Herot and Guy Weinzapfel, "One-Point Touch Input of Vector Information for Computer Displays," in *Proceedings of the 5th Annual Conference on Computer Graphics and Interactive Techniques*, SIGGRAPH '78 (New York: ACM, 1978), 210–216.

78 Weinzapfel and Negroponte, "Architecture-by-Yourself," 74–75.

79 Ibid., 74.

80 Negroponte and Groisser, "URBAN5: A Machine that Discusses Urban Design," 113.

81 Ibid.

82 Stanford Anderson, ed., *Planning for Diversity and Choice: Possible Futures and Their Relations to the Man-Controlled Environment* (Cambridge, MA: MIT Press, 1968). The book was based on a conference held at Endicott House in Dedham, Massachusetts on 13–16 October 1966 with the title Inventing the Future Environment. The initiative was sponsored by the Graham Foundation, the American Institute of Architects-Princeton Educational Research Project, and the Department of Architecture at MIT.

83 Examples of projects and committees focusing on future studies were the Ford Foundation-funded Futuribles project initiated under the directorship of Bertrand de Jouvenel in 1963; The Committee for the Next Thirty Years directed by Michael Young and Mark Abrams, as part of the English Social Science Research Council; and the Commission on the Year 2000 of the American Academy of Arts and Sciences, initiated in 1965 with Daniel Bell as director. For a list of initiatives organized around the category of the "future" see Bettina J. Huber, "Studies of the Future: A Selected and Annotated Bibliography," in *Sociology of the Future: Theory, Cases and Annotated Bibliography*, edited by Wendell Bell and James Wau (New York: Russell Sage Foundation, 1971), 339–469. For a critical history of futurology see Jenny Andersson, *The Future of the World: Futurology, Futurists, and the Struggle for the Post Cold War Imagination* (Oxford: Oxford University Press, 2018).

84 Nicholas Negroponte, "Limits to the Embodiment of Basic Design Theories," in *Basic Questions of Design Theory*, edited by William R. Spillers (Amsterdam: North-Holland, 1974), 61.

85 Groisser and Negroponte, *Computer Aids to Participatory Architecture*, i.

86 Ibid.

87 Ibid., 2.

88 Negroponte and Groisser, "URBAN5: A Machine that Discusses Urban Design," 107.

89 Groisser and Negroponte, *Computer Aids to Participatory Architecture*, 2.

90 Ibid., 73.

91 Ibid., 72.

92 Ibid., 23.

93 Negroponte, *Soft Architecture Machines*, 119.

94 Groisser and Negroponte, *Computer Aids to Participatory Architecture*, 27.

95 In the computer graphics community assembled around the CAD Project at MIT, "sketching" stood for a fluid, dynamic interaction between user and machine. A register of ambiguity and typically associated with the early stages of the design process, sketching symbolized that which was tentative and changing. Sketching formed a slogan and impetus for new developments both in graphical interfaces (light pens, tablets, point-digitizers, and other devices) that would make drawing in a computer feel and look more like sketching, but also in computer graphics and machine vision that would enable computers to "read" and handle their users' sketches. See Nicholas Negroponte, "Sketching: A Computational Paradigm for Personalized Searching," *JAE* 29, no. 2 (1975): 26–29.

96 Negroponte, *Soft Architecture Machines*, 65. On the Sylvania data tablet see James F. Teixeira and Roy P. Sallen, "The Sylvania Data Tablet: A New Approach to Graphic Data Input," in *Proceedings of the April 30–May 2, 1968, Spring Joint Computer Conference* (New York: Association for Computing Machinery, 1968), 315–321.

97 Negroponte, *Soft Architecture Machines*, 64.

98 Ibid.

99 Ibid., 65.

100 Groisser and Negroponte, *Computer Aids to Participatory Architecture*, 28.

101 Ibid.

102 Ibid., 19.

103 Ibid.

104 Ibid., 7. Negroponte drew from a tripartite formulation based on British cybernetician and frequent Architecture Machine visitor Gordon Pask's "conversation theory," which advanced a cybernetic approach to information exchange between two entities, to cognition, and to learning. A successful interaction with a computer, he argued, required "(1) its model of

you, (2) its model of your model of it and (3) its model of your model of its model of you." See Negroponte, *Soft Architecture Machines*, 60.

105 Ibid., 9–10.

106 Philip Steadman, *Architectural Morphology* (London: Pion, 1983).

107 Ibid., 144.

108 Ibid.

109 Ibid.

110 Ibid., n.p.

111 Philip Steadman, "Graph Theoretic Representation of Architectural Arrangement," *Architectural Research and Teaching* 2 (1973): 161–172; Philip Steadman, "Graph-Theoretic Representation of Architectural Arrangement," in *The Architecture of Form*, edited by Lionel March (London: Cambridge University Press, 1976), 94–115.

112 Lionel March, "Modern Movement to Vitruvius: Themes of Education and Research," *Royal Institute of British Architects Journal* 81, no. 3 (1972): 103.

113 Steadman, "Graph-Theoretic Representation of Architectural Arrangement," 103.

114 Ibid.

115 Ibid.

116 W. J. Mitchell, Philip Steadman, and Robin S. Liggett, "Synthesis and Optimization of Small Rectangular Floor Plans," *Environment and Planning B: Planning and Design* 3, no. 1 (1976): 37–70.

117 Ibid., 38.

118 Ibid., 41.

119 Ibid., 42.

120 Ibid., 53–54.

121 Ibid., 55.

122 Ibid.

123 Lionel March and Philip Steadman, "From Descriptive Geometry to Configurational Engineering," in *Proceedings. International Conference on Descriptive Geometry*, edited by G. K. Hilliard (Washington, DC: American Society for Engineering Education, 1978), 21.

124 Ibid., 24.

125 Ibid., 22.

126 Centre for Configurational Studies, "Triennial Report 1978–81," 1981, Open University Archive.

127 Ibid.

128 Scott Campbell, "'Wat for Ever': Student-Oriented Computing at the University of Waterloo," *IEEE Annals of the History of Computing* 35, no. 1 (2013): 11–22.

129 "A Brief History on the Faculty of Mathematics," University of Waterloo, https://uwaterloo.ca/math/about/our-history

130 "Remembering George Soulis, Father of Systems Design Engineering," *University of Waterloo Daily Bulletin*, February 1, 2018, https://uwaterloo.ca/daily-bulletin/2018-02-01

131 Peter Seitz and Martin Krampen, eds., *Design and Planning 2: Computers in Design and Communication* (New York: Hastings House, 1967).

132 Peter Seitz, "Design and the Computer," *Design Quarterly* (Walker Art Center, Minneapolis) 66/67 (1967).

133 Lionel March and Ray Matela, "The Animals of Architecture: Some Census Results on N-omino Populations for $N = 6, 7, 8$," *Environment and Planning B* 1 (1974): 193–216.

134 Ibid., 193.

135 Ibid.

136 R. S. Frew, R. K. Ragade, and P. H. Roe, "The Animals of Architecture," in *Proceedings of the EDRA 3/AR 8 Conference*, edited by William J. Mitchell (Los Angeles: School of Architecture and Urban Planning, University of California, Los Angeles, 1972), 23-2-1–23-2-7.

137 William J. Mitchell, introduction to *Proceedings of the EDRA 3/AR 8 Conference*.

138 Frew, Ragade, and Roe, "The Animals of Architecture," 23-2-1.

139 Ibid., 23-2-6.

140 March and Matela, "The Animals of Architecture," 197.

141 Ibid., 200.

142 Ibid., 212.

143 Ibid.

144 Ibid.

145 William J. Mitchell and Robert Dillon, "A Polyomino Assembly Procedure for Architectural Floor Planning," in *Proceedings of the EDRA 3/AR 8 Conference*, 23-5-1–23-5-12.

146 Ibid., 23-5-1.

147 March and Matela, "The Animals of Architecture," 196.

148 Centre for Configurational Studies, "Triennial Report 1978–81," 16.
149 Ibid., 22.
150 Ibid., 19.
151 Lionel March and Chris Earl, "On Counting Architectural Plans," *Environment and Planning B* 4 (1977): 58.
152 The interlude rehearses the lineage of engagement with polyhedra proposed in ibid., 61.
153 Ibid., 57.
154 Ibid.
155 Ibid., 59 (emphases in the original).
156 Ibid., 60.
157 Ibid., 61.
158 Ibid.
159 Ibid., 59.
160 Ibid.
161 Ibid.
162 Ibid.
163 Ibid., 62.
164 Ibid., 78.
165 Ibid.
166 Ibid., 75.
167 Ibid., 78.
168 Ibid., 79.
169 George Stiny, "Two Exercises in Formal Composition," *Environment and Planning B: Planning and Design* 3, no. 2 (1976): 187–210.
170 Ibid., 194.
171 Ibid.
172 Ibid., 209.
173 Centre for Configurational Studies, "Triennial Report 1978–81," 16.
174 Ibid.
175 C. J. Bloch and Ramesh Krishnamurti, "The Counting of Rectangular Dissections," *Environment and Planning B* 5 (1978): 207–214. Krishnamurti also presented a computer algorithm for shape grammars in Ramesh Krishnamurti, "The Construction of Shapes," *Environment and Planning B: Planning and Design* 8, no. 1 (1981): 5–40.
176 Noam Chomsky, *Syntactic Structures* (The Hague: Mouton, 1957).
177 George Stiny and James Gips, "Formalization of Analysis in Design and the Arts," in Spillers, *Basic Questions of Design Theory*, 507–530.
178 Ibid.
179 Ibid., 507.
180 Ibid., 508.
181 George Stiny, "Letter to the Editor: Spatial Relations and Grammars," *Environment and Planning B* 9 (1982): 113–114.
182 Ibid., 113.
183 Ibid., 114.
184 Christopher F. Earl, "A Note on the Generation of Rectangular Dissections," *Environment and Planning B: Planning and Design* 4, no. 2 (1977): 241.
185 "Introduction: Computer-Aided Participatory Design," in Negroponte, *Soft Architecture Machines*, 95.
186 Ibid., 96.
187 Ibid.
188 Ibid.
189 Negroponte, *Soft Architecture Machines*, 115.
190 Ibid.
191 Ibid.
192 Ibid.

Chapter 5

1 See landmark texts on digital architecture: Mario Carpo, *The Digital Turn in Architecture 1992–2012* (Chichester, UK: Wiley, 2013); Antoine Picon, *Digital Culture in Architecture: An Introduction for the Design Professions* (Basel: Birkhäuser, 2010); Mario Carpo, *The Second Digital Turn: Design beyond Intelligence* (Cambridge, MA: MIT Press, 2017). On digital architecture as a historical phenomenon see Andrew Goodhouse, ed., *When Is the Digital in Architecture?* (Montreal: Canadian Centre for Architecture, 2017); and Nathalie Bredella, *The Architectural Imagination at the Digital Turn* (Abingdon, UK: Routledge, 2022). See also the debate: Martin Bressani, Mario Carpo, Reinhold Martin, and Theodora Vardouli, edited by Antoine Picon, "L'architecture à l'heure du numérique," in *Perspective: Actualité en histoire de l'art* (Paris: Institut National d'Histoire de l'Art, 2019), 113–140.
2 Smoothness and surface continuity as the defining characteristic of digital architecture

have been a stable object of counterdefinitions of digital architecture. See for example, Mario Carpo, "Digital Style," *Log* 23 (2011): 41–52; Gilles Retsin, ed., "Discrete: Reappraising the Digital in Architecture," *Architectural Design*, no. 258 (2019).

3 Claude Berge, *The Theory of Graphs and Its Applications* (New York: Methuen, 1962), ix.

4 On ghosts and the notion of haunting by futures past, see Jacques Derrida's classic work *Specters of Marx: The State of the Debt, the Work of Mourning, and the New International* (London: Routledge, 1994), and the field of "hauntology."

5 Philipp Von Hilgers, "The History of the Black Box: The Clash of a Thing and Its Concept," *Cultural Politics* 7, no. 1 (2011): 41–58.

6 Andrew Witt, "Grayboxing," *Log* 43 (2018): 69–77.

7 Jonathan Barnett, "Glass Box and Black Box," *Architectural Record* (1968): 127–128.

8 Christopher J. Jones, *Design Methods: Seeds of Human Futures* (London: Wiley-Interscience, 1970).

9 Barnett, "Glass Box and Black Box," 128.

10 Ibid.

11 Andrew Pickering, "Cybernetics and the Mangle: Ashby, Beer and Pask," *Social Studies of Science* 32, no. 3 (2002): 413–437.

12 For a compelling reflection on the spatiality of the closet see Michael Brown, "Where Is the Closet?," in *Closet Space: Geographies of Metaphor from the Body to the Globe* (London: Routledge, 2000), 141–149. The work draws from the foundational theorization of the "closet" in queer studies in Eve Kosofsky's classic work *Epistemology of the Closet* (Berkeley: University of California Press, 1990).

13 Friedrich Nietzsche, "Nietzsche contra Wagner," in *The Works of Friedrich Nietzsche*, translated by Thomas Common, vol. 3, *The Case of Wagner* (London: T. Fisher Unwin, 1899), 93.

Index

Page numbers followed by "f" indicate figures.

Activity Data Method, 110
Advanced Research Projects Agency (ARPA), 149, 208n46, 215n77
Albert A. Hoover & Associates, 113
Alexander, Christopher
  The Atoms of Environmental Structure, 110–112, 118
  Center for Environmental Structure and, 116–118, 124, 134, 136, 196, 212n10
  Chermayeff and, 14–15, 104, 107
  "A City Is Not a Tree," 117, 129
  Community and Privacy, 143
  Conference on Design Methods and, 87
  "The Coordination of the Urban Rule System," 110
  data and, 84–87
  Environmental Form Incorporated and, 114
  failure and, 103, 104f
  generativity and, 124, 128–130
  Hierarchical Decomposition System (HIDECS) II, 106, 109, 117
  Hierarchical Decomposition System (HIDECS) III, 117
  Ishikawa and, 107, 109, 111–112, 116–118, 124, 134
  layouts and, 87, 103–104, 105f
  modern mathematics and, 112
  Notes on the Synthesis of Form, 85–86, 106, 111–112
  Offices Development Group, 110
  participatory design and, 134–137
  A Pattern Language, 107, 109, 122, 128, 133
  pattern languages and, 116–128
  "Relational Complexes in Architecture," 112
  relational theory and, 110–111
  "The Revolution Finished Twenty Years Ago," 85
  Silverstein and, 107, 116–118, 122, 124, 134
  "Systems Generating Systems," 124
  "Ten Year Program," 109
  at University of California, Berkeley, 107–117, 122–124
  "The Urban House," 104
  W-algorithm and, 125f–127f
Algebra
  Boolean, 24, 179
  geometry and, 35, 164, 179
  textbooks and, 24, 27, 35–36, 39
Algorithms
  automation and, 10, 79–82, 87, 94, 101, 128–129, 178, 194
  CB, 178
  data and, 82, 85, 87
  layouts and, 91, 93–94, 101, 103
  patterns and, 106–107, 127f
  RK, 178
  W, 125f–127f
American Bureau of Standards, 116
American Institute of Architects (AIA), 164
American Mathematical Monthly, 162
ARCH+ (magazine), 58
Archer, Bruce, 89–90, 91f
Archigram group, 48, 74
Architects' Journal, The, 81, 101, 148
Architects' Journal Information Library, 88, 90–91
Architectural Design, 15–16, 24, 27, 97, *143*
Architectural Forum, 117
Architectural Morphology (Steadman), 155–156
Architecture and Building Aids Computer Unit (ABACUS), 97
Architecture and the Computer (conference), 103
Architecture-by-Yourself, 150–153, 155
Architecture d'aujourd'hui, 67
Architecture Machine
  ARPA and, 215n77
  Brodey and, 150

Architecture Machine (*continued*)
Burnham and, 149
evolutionary learning and, 149, 152, 181
Friedman and, 150–151, 155–156, 181–182, 187, 189
Graham Foundation and, 149
Groisser and, 153–155
HUNCH and, 154
intelligent environments and, 149–150
National Science Foundation, 150, 153
Negroponte and, 148–156, 181–182, 187, 189–190, 215n77, 216n104
participatory design and, 151–154, 190
Pask and, 150, 216n104
SQUINT and, 154
URBAN5 and, 149, 152–153, 214n59
Weinzapfel and, 150–152, 215n74
YONA and, 138, 150–152, 182
Architecture Mobile (Friedman), 57
Architecture of Form, The (March), 75
Asimow, Morris, 86
"Aspects of Living in an Architecture Machine" (Negroponte), 149
Atlas computer, 88
Atoms of Environmental Structure, The (Alexander and Poyner), 110–112, 118
"Automatic Generation of Minimum-Standard House Plans, The" (Steadman), 46
Automation
algorithms and, 10, 79–82, 87, 94, 101, 128–129, 178, 194
data and, 82, 87
layouts and, 94, 101, 103
patterns and, 110
tools and, 79–82, 87, 94, 101, 103, 110, 128–130, 206nn3–5, 208n45

Babbage, Dennis, 19
Baker, Michael, 109, 112
Banham, Reyner, 13, 148
Barfield, Owen, 183
Barnett, Jonathan, 189
Barr, Alfred, 56
Bartlett School of Architecture, 30
Batty, Michael, 162
Bauhaus, 7, 13–14, 19, 86, 98
Bauwelt, 51
Bay Area Rapid Transit (BART), 108–110, 112, 116–117
Bell Labs, 152
Bense, Max, 57
Birkhoff, George David, 57
Black boxes, 188–190
Blake, William, 122
Bloch, Cecil, 178
Bofill, Ricardo, 48, 74
Bolt Beranek and Newman (BBN), 208n46
Boolean algebra, 24, 179
Boston Architectural Center, 103
Boulding, Kenneth, 24, 187, 202n43
Bourbaki, Nicolas, 7–8, 35
Brand, Stewart, 133
British Architectural Students Association (BASA), 30, 81, 148
British Constructionist Group, 56
British Productivity Council, 81
Broadbent, Geoffrey, 86–87, 129, 148
Brodey, Warren, 150, 190
Brooks, R. L., 46
Building Research Board, 101
Building Research Institute, 14
Building Research Station (BRS), 81, 89, 109

Calculus, 76, 140
Callenbach, Ernest, 133
Cambridge Mathematical Laboratory, 46–47
Cambridge Review, 19
Carnegie Institute of Technology, 143
Cascade
multi-service centers and, 117–118, 119f–120f, 124, 211n135
patterns and, 118, 119f–120f, 124, 128–129, 136, 140, 180
Case Institute of Technology, 86
CB algorithm, 178
Center for Environmental Structure
Alexander and, 116–118, 124, 134, 136, 196, 212n10
background of, 113–115
founding of, 116
Ishikawa and, 107, 116, 117–118
Montgomery on, 116
multi-service centers and, 118–121
National Institute of Mental Health and, 124
participatory design and, 134–137
pattern language and, 116–128
Silverstein and, 107, 116, 117–118
Centre for Configurational Studies
Earl and, 166
enumeration and, 172–175
founding of, 161
Krishnamurti and, 178
March and, 161–162
research areas of, 161
Science Research Council and, 166, 178
shape grammars and, 178
Stiny and, 178
Chamberlain, Neville, 81
Chermayeff, Serge
Alexander and, 14–15, 104, 107
CIAM and, 14, 200n11
Harvard-MIT Joint Center for Urban Studies, 15
patterns and, 106–108, 123f
"The Urban House," 104
Chomsky, Noam, 178
Circle: International Survey of Connectionist Art, 19
"City Is Not a Semilattice Either, A" (Harary and Rockey), 129
"City Is Not a Tree, A" (Alexander), 117, 129
Civil rights, 118, 133
Colbert, Charles, 13–14
Cold War, 29, 199n24, 203n65
Columbia University, 13
Combinatorics
Architecture Machine and, 181–182, 189
Bourbaki and, 8
enumeration and, 156, 180–181
FLATWRITER and, 146, 155, 189
Friedman and, 53, 180–182
March and, 161, 162–165, 182
pattern and, 117, 124, 180
possibility and, 46, 130, 173–174, 176, 179
spatial urbanism and, 53, 67
Steadman and, 46, 161, 180
Stiny and, 176
Waterloo school of, 173–174
Committee on the Application of Computers in the Construction Industry (CACCI), 97
Community and Privacy (Alexander and Chermayeff), 143

Computer-Aided Design (journal), 151
"Computer Aids to Participatory Architecture" (Negroponte and Groisser), 154
Computers
Atlas, 88
Bolt Beranek and Newman (BBN) and, 208n46
Ferranti, 88
IBM 360/91, 157–158
IBM 370/155, 157
IBM 709, 106
IBM 2250, 214n59
IMAGE and, 215n74
IMLAC PDS-1D, 187
KDF9 English Electric, 82, 91, 94
PDP-10, 166
VAX 11/780, 161
"Computers in Building" series, 88–89
Conference on Systematic and Intuitive Methods in Engineering, Industrial Design, Architecture and Communications (Conference on Design Methods), 86–87, 89, 137, 148
Configuration, 90, 111, 140–141, 143, 146, 149, 151, 155, 159, 161, 163, 180–181
Congrès Internationaux d'Architecture Moderne (CIAM)
demise of, 7, 13
founding agenda of, 14
Friedman and, 51
Giedion and, 198n19, 202n34
Le Corbusier and, 198n19, 202n34
Mumford and, 200n11
"Coordination of the Urban Rule System, The" (Alexander), 110
Coplanner, 208n46
CPM (critical path method), 88
CRAFT, 208n47
Croce, Benedetto, 25
Cross, Nigel, 148

Data
Alexander and, 84–87
algorithms and, 89–94
automation and, 82, 87
categories and, 84–87, 104
diagrams and, 82, 84f, 87
failure and, 104–105
hospitals and, 89–94
layouts and, 82, 98, 101–103
tools and, 80–87
DATA (Hill), 56–57
Davis, Gerald, 113–114
De Carlo, Giancarlo, 152
Department of Defense (US), 149, 208n46
Descartes, Blanche (pseud.), 1, 7–8, 37, 46, 57, 162
Design Integration Ltd., 89
Design Methods Group
EDRA 3 and, 164
first conference of, 149, 215n76
founding event of, 162
Negroponte and, 149–150, 152
Design Methods: Seeds of Human Futures (Jones), 189
Design Quarterly, 162
Design Research Society, 86, 148
De Stijl, 86
Devin House, 42, 45f
Diagrams, 5
architectural theory and, 6
layouts and, 89f, 91, 93–94, 98–99, 103
patterns and, 106, 109–110, 113–117, 120f
string, 82, 91–94, 98–99, 103, 129
textbooks and, 27, 36, 43
Dickens, Peter, 16
Dieudonné, Jean, 35
Dillon, Robert, 165
DOMINO, 165–166
Doshi, Balkarishna V., 84, 137, 140
Dürer, Albrecht, 172

Earl, Christopher, 166, 172–174, 180
Echenique, Marcial, 16
École Nationale Supérieure des Beaux-Arts, 74
Ecotopia (Callenbach), 133
Edgar J. Kaufmann Foundation, 116
EDSACs, 46–47
Education Act, 81
Effort, 143, 144f–145f, 181
"Eight Homes for Modern Living" (Wright), 23–24
Eldars, Mohamed Zakaria Ahmed
layouts and, 91–94, 98–103
Whitehead and, 91–94, 98–103, 110–111, 156, 165, 208n47
Emmerich, David Georges, 51
Enumeration, 156, 180
Environmental Design Research Association (EDRA), 164–166
Environmental Form Incorporated (EFI), 114–116
Environmental Geometry (Nuffield Mathematics Project), 36
Environment and Planning B (journal), 162–166, 172, 176, 179
Erlangen Program, 35, 40
Euclid, 4, 8, 25, 35, 203n74
Euler, Leonhard, 1, 56, 170
"Expanding Universe, The" (Blanche Descartes), 1

Facheux, Claude, 70
Fair Housing Act, 133
Farr, Michael, 89
Ferranti Ltd., 89–90
Ferranti Research Center, 88
FLATWRITER
combinatorics and, 146, 155, 189
Friedman and, 140–148, 155, 181, 189
geometry and, 140–141
infrastructures and, 140–143, 146, 148, 155, 181, 189
process of, 143–148
Floor plans
activity and, 82, 91–94, 110
generation of, 82, 91–94, 130, 156–158
menu of, 53, 141, 172–175
representation of, 46–47, 53, 82, 98, 172–175
Ford Foundation, 159, 214n71, 216n83
Form
Alexander and, 19, 108, 109, 112, 114
isomorphism and, 20, 22, 24
March and, 16–24
Martin and, 16, 19, 24
science of, 163, 165, 176
Steadman and, 23f, 156
Wright and, 20, 22
"Formalization of Analysis and Design in the Arts" (Stiny and Gips), 178–179
Fortran Monitor system, 106
Foundations of Geometry (Hilbert), 8
"Four Great Makers and the Next Phase in Architecture, The" (event), 13–15

Frankl, Paul, 176
Frew, Robert, 164
Friedman, Yona
Architecture Machine and, 150–151, 155–156, 181–182, 187, 189
CIAM and, 51
combinatorics and, 180–182
exhaustive enumeration and, 180–181
FLATWRITER and, 140–148, 155, 181, 189
Harary and, 56–58, 129, 181
Hill and, 56–61, 62f, 64f, 66f, 129, 181
isomorphism and, 74–75
possibility and, 155–156
Pour une architecture scientifique, 48, 51–52, 58, 67, 74, 140–141, 150
Ragon on, 146
Ville Spatiale, 51–53, 67, 74–75, 143, 146
YONA and, 138, 150–152, 181, 182
Future of Architecture, The (Wright), 173

Gabo, Naum, 19
Gardner, Martin, 46
Geometry
algebra and, 35, 164, 179
axiomatic approach and, 203n76
combinatorics and, 185–186
descriptive, 159, 161
Euclidean, 4, 8, 25, 35, 203n74
Hilbert on, 8
isomorphism and, 75–76
Klein on, 35, 179
layouts and, 94, 98, 100f
new mathematics and, 25–46
patterns and, 111
peril and, 13–15
transformation and, 25, 32, 35–36, 40, 161, 179
Geometry of Environment, The (March and Steadman), 2f
"Electrical Networks and Mosaics of Rectangles," 43–48
isomorphism and, 76
layouts and, 98
"Mappings and Transformations," 39
"Planar Graphs and Relations," 40–42
textbooks and, 25–29, 36, 39, 43–48
Giedion, Sigfried, 202n34
Gilbert, Stephen, 56
Ginsberg, Allen, 122
Gips, James, 178–179
Glass box, 188–190
Golomb, Solomon, 162
Goodman, Robert, 152
Grabow, Steven, 136
Graham, Wes, 162
Graham Foundation, 159
Grands ensembles, 146
"Graph Theoretic Representation of Architectural Arrangement" (Steadman), 156
Graph Theory (Harary), 172
Grason, John, 215n76
Gray, Crispin, 47
Greibach, Sheila, 178
Groisser, Leon, 153–155
Gropius, Walter, 13–14, 58, 202n34
Groupe d'Études d'Architecture Mobile (GEAM), 51, 67, 137, 146
Groupe International d'Architecture Prospective (GIAP), 67
Günschel, Günter, 51

Habitat évolutif, 146
Habraken, N. John, 137, 140
Harary, Frank
aesthetics and, 70
Alexander and, 129
Friedman and, 56–58, 129, 181
graphs, 5, 39, 56, 70, 129, 172, 178
Hill and, 57
isomorphism and, 5, 74
models and, 39
Harvard-MIT Joint Center for Urban Studies, 15, 20, 104, 106
Hawkes, Dean, 75, 97
Hervé, Lucien, 50f
HIDECS (HIerarchical DECOmposition System), 106, 109, 117
Hilbert, David, 8, 35, 203n79
Hill, Anthony
aesthetics and, 56–57
Friedman and, 56–61, 62f, 64f, 66f, 129, 181
Harary and, 56
Hirshen, Sandy, 117
Hochschule für Gestaltung, Ulm, 89
Hollein, Hans, 74
Homes for Today and Tomorrow (Parker Morris Committee), 46
Hospitals
Coplanner and, 208n46
data and, 89–94
layouts and, 88–94, 97–101, 208n46
Nuffield Trust and, 93
"House for the Modern Family, The," 14, 200n6
Houses Generated by Patterns (guide), 134
Huet, Bernard, 74
Humanism, 37, 182
Human Resources Administration (HRA), 118
HUNCH, 154, 182
Hunts Point, 117–118
Huxtable, Ada Louise, 117
Hyslop, Patrick, 109, 112

IBIS, 212n6
IBM 360/91, 157–158
IBM 370/155, 157
IBM 709, 106
IBM 2250, 214n59
Iconography, 5–6, 14–15, 76, 181, 191
IMAGE, 151, 153–154, 215n74
IMLAC PDS-1D, 187
Immutability, 3, 53, 130, 212n155
Imperial College, 137
Imperial Hotel, Tokyo, 48
Information aesthetics, 57, 205n137
Infrastructures
change and, 150–155
choice and, 140–150
combinatorics and, 179–183
control and, 133–140
FLATWRITER and, 140–143, 146, 148, 155, 181, 189
patterns and, 133–136, 140–143, 155, 161, 164, 166, 170, 180, 212n10
politics and, 133–136, 140, 156, 180, 213n42, 215n76
possibility and, 155–170
shape grammars and, 178–179, 183, 193, 218n175
spatial urbanism and, 67
topology and, 154, 157, 161, 164, 166, 174, 182
Interdata Model 3, 149
International and Universal Exposition (Expo '67), 162
International Conference on Descriptive Geometry, 159, 161

International Design Conference, 84, 212n17
International Dialogue of Experimental Architecture (IDEA), 74
International Society for Geometry and Graphics, 161
Introduction to Design (Asimow), 86
Ishikawa, Sara, 210n83
Alexander and, 107, 109, 111–112, 116–118, 119f–120f, 124, 134
BART and, 109, 112, 116, 118
entrance relations and, 211n106
Isomorphism
Alexander and, 74
form and, 20, 22, 24
Friedman and, 74–75
Harary and, 74
Land Use Built Form Studies (LUBFS) and, 75–76
layouts and, 97, 100f
March and, 24, 75
Steadman and, 75
textbooks and, 29
unpacking of term, 10

Japanese teahouse, 136
Johnson, Timothy, 150–151
Jones, John Christopher, 87, 148, 161, 189

Kant, Immanuel, 1
Kasabov, George, 30
Kaufmann International Design Award, 117
KDF9 English Electric Computer, 82, 91, 94
Kesavan, H. K., 162
Kikutake, Kiyonori, 137
Kirchhoff's second law, 46
Klein, Alexander, 98
Klein, Felix, 35–36, 40, 179, 203n79
Kőnig, Dénes, 70
Königsberg bridge problem, 1, 2f
Krampen, Martin, 162
Krishnamurti, Ramesh, 178, 218n175
Kuratowski, Kazimierz, 43
Kurokawa, Kisho, 48

Landscape (journal), 103, 137
Land Use Built Form Studies (LUBFS)
forms and, 15–16, 19–20, 24–25
founding of, 16–19
funding of, 201n33
isomorphism and, 75–76
layouts and, 97–98, 99f
manifesto of, 15–16, 24
mathematical models and, 15–16, 20, 97–98
textbooks and, 27, 36, 46
Larson, Theodore, 104
Layouts
Alexander and, 103–104
algorithms and, 91, 93–94, 101, 103
computers and, 87–94, 97, 101, 103
diagrams and, 83, 91, 93–94, 98–99, 103
Eldars and, 91–94, 98–103
graphs and, 98–99, 101
hospitals and, 88–94, 97–101, 208n46
Levin and, 101
Muther and, 93–94
pattern and, 97–98, 101, 103
Steadman and, 46–47, 158
Tabor and, 97–101
Whitehead and, 90–94, 98–101, 103, 208n47
Le Corbusier, 13, 40, 48, 198n19, 199n25, 202n34
Leibniz, Gottfried Wilhelm, 56
Leonardo (journal), 57
Levin, Peter, 101
Lévi-Strauss, Claude, 8
Licklider, J. C. R., 208n46
Life magazine, 23–24
Liggett, Robin, 157
Lindheim, Roslyn, 116
Lönberg-Holm, Knud, 104

Machine Recognition and Inference Making in Computer Aids to Design, 150
Malina, Frank, 57
Manheim, Marvin, 106, 116
March, Lionel
Center for Configurational Studies and, 161–162
form and, 16–24
The Geometry of Environment, 2f, 25–29, 36, 39, 43–48, 76, 98, 159, 161
isomorphism and, 24, 75
Matela and, 162–166, 176
polyominoes and, 162–166, 167f–169f
possibility and, 155–156, 159, 161–176
Whitehall and, 18f, 19
Martin, Kenneth, 56
Martin, Leslie, 16, 19
Oxford Conference and, 80–81, 84
Scruton on, 75
textbooks and, 29
Martin, Mary, 56
Martin Centre for Architectural and Urban Studies, 33f, 48f, 99f, 178
Massachusetts Institute of Technology (MIT)
Civil Engineering Systems Laboratory, 106
Computation Center, 106
Computer-Aided Design Project, 214n71
Department of Architecture, 138, 151
School of Architecture and Planning, 150–151
Matela, Ray, 162–166, 176
Matrices
frameworks and, 53, 54f, 70
layouts and, 93–94
possibility and, 157, 164
relationship, 93–94
textbooks and, 25, 27, 32, 35–36, 39, 47
Maymont, Paul, 51
Melencolia I (Dürer), 172
Melp! (magazine), 74
Menu, 140–141, 146, 182
Meyer, Hannes, 98
Mies van der Rohe, Ludwig, 13
Ministry of Health, 89
Ministry of Housing and Local Government, 46
Ministry of Public Building and Works, 97, 109–110
Mitchell, William, 75, 157, 165
Modernism
architectural, 6–7, 9, 13–14, 16, 76, 200n11
graphs and, 6–9, 187
mathematical, 6–9, 76
peril and, 13, 16
second, 7, 199n23
style and, 6, 13

Moles, Abraham, 57
Montgomery, Roger, 85, 116
Moore, Ian, 110
Mühlestein, Erwin, 48
Multi-service centers, 117–118, 119f–120f, 124, 211n135
Mumford, Eric, 200n11
Muther, Richard, 93–94

National Health Service Act, 81
National Health System, 89
National Institute of Mental Health, 116, 124, 134
National Productivity Year, 81
National Science Foundation, 148–150, 153
Negroponte, Nicholas
  Architecture Machine and, 148–156, 181–182, 187, 189–190, 215n77, 216n104
  "Aspects of Living in an Architecture Machine," 149
  Brodey and, 150
  "Computer Aids to Participatory Architecture," 154
  computers and, 139f, 148, 151–154, 181–182, 216n104
  Friedman and, 150
  Groisser and, 153–155
  HUNCH and, 154, 182
  Noll and, 152
  Pask and, 216n104
  SEEK and, 149
  Soft Architecture Machines, 181
  SQUINT and, 154, 182
  URBAN5 and, 149–150, 152, 189, 214n59
  Weinzapfel and, 150–152
Networks
  frameworks and, 53, 56, 58, 66f, 68f
  graphs and, 2, 9, 58, 87
  isomorphisms and, 76
  neural, 76, 186
  patterns and, 128–129
  possibility and, 155, 162, 166
  researcher, 53, 56, 155, 162
  skeletons and, 185–186, 190
  textbooks and, 25, 36, 43
Newman, William, 19
New mathematics
  design methods and, 86–87
  geometry and, 35–38
  graph theory and, 36–37
  isomorphism and, 75–76
  patterns and, 112
  politics and, 203n65
  possibility and, 159, 161, 179
  Quadling and, 32, 35
  School Mathematics Project and, 32
  textbooks and, 27, 29, 32, 35–39, 203n51
  Thwaites and, 32, 35
New Towns Act, 81
Nicholson, Ben, 19
Noll, Michael, 152
Notes on the Synthesis of Form (Alexander), 85–86, 106, 111–112
Nuffield Foundation, 81, 89
Nuffield Mathematics Project, 30, 31f, 36
Nuffield Provincial Hospitals Trust, 93

Office for Naval Research, 149
Offices Development Group, 109–110, 112
Open University, 161, 166, 178
Operational graphs, 99, 129–130
Oregon Experiment (Center for Environmental Structure), 134
Organizational graphs, 99
Ornamentation, 172–173, 183
Osaka, Expo 70, 140, 212n10
Otto, Frei, 51
Oxford Conference, 81

Paragrams, 56
Parker Morris Committee, 46
Pask, Gordon, 182, 190, 216n104
Pasmore, Victor, 56
Pattern Language, A (Alexander et al.)
  Brand on, 133
  control and, 133, 136, 140
  development of, 107–110, 116–122, 124, 128–130
  participatory design and, 133–134, 136, 140
Patterns
  Alexander and, 106–128
  algorithms and, 106–107, 127f
  cascades and, 118, 119f–120f, 124, 128–129, 136, 140, 180
  Chermayeff and, 106–108, 123f
  computers and, 106, 110, 112–113, 117
  diagrams and, 106, 109–110, 113–117, 120f
  hospitals and, 111
  infrastructures and, 133–136, 140–143, 155, 161, 164, 166, 170, 180, 212n10
  Japanese teahouse and, 136
  layouts and, 97–98, 101, 103
  new mathematics and, 112
  possibility and, 161, 164, 166, 170
  systems and, 106–110, 124
  Tabor and, 110–111
  tree structures and, 9, 117
  University of California, Berkeley and, 107–117, 122–124
  Whitehead and, 110–111
PDP-10, 166
Peirce, Charles Sanders, 56
Peril, 13–15
Permutations, 68f, 189, 208n47
PERT (program evaluation research techniques), 88
Picabia, Francis, 56
Planar graphs 25, 40, 43, 45f, 52–53, 54f, 68f, 158, 162–164, 172–174
Planning for Diversity and Choice (Anderson), 152
Polyhedra, 57, 62f, 170, 172, 174, 218n152
Polyominoes
  Golomb and, 162
  March and, 162–166, 167f–169f
  possibility and, 162–166, 167f–169f, 174, 176, 182
Possibility
  algorithms and, 157–158, 178
  Architecture Machine and, 155–156
  automation and, 156, 165, 178
  computers and, 155–157, 161–162, 165–166, 178–179
  Friedman and, 155–156
  geometry and, 156, 159, 161, 164–165, 166f–169f, 172–173, 179
  graphs and, 158, 162–164, 172–174
  Harary and, 172, 178
  hospitals and, 166
  infrastructures and, 155–170
  Land Use Built Form Studies (LUBFS) and, 178
  mapping and, 163, 166, 174, 176
  March and, 155–156, 159, 161–176
  pattern and, 161, 164, 166, 170
  shape grammars and, 178–179, 183, 193, 218n175

Steadman and, 155–161, 166
transformation and, 161, 173, 178–179
Poststructuralism, 6
Pour une architecture scientifique (Friedman), 48, 51–52, 58, 67, 74, 140–141, 150
Poyner, Barry, 110–112
Prefabrication, 58, 137, 141, 156, 181
Price, Cedric, 74
"Progress and Harmony for Mankind" (Osaka Expo), 140
Proyecto Experimental de Vivienda (Lima), 134

Quadling, Douglas, 32, 35–36

Ragade, Rammohan, 164
Ragon, Michel, 67, 146
RAND Corporation, 114
Read, R. C., 162
Rectangular dissections, 46, 157–158, 178, 182
"Reflections on the Architecture of the Future: Criteria for Town-Planning" (Friedman), 56
"Relational Complexes in Architecture" (Alexander et al.), 112
Relational theory, 110–111
Relationships matrix, 93–94
"Revolution Finished Twenty Years Ago, The" (Alexander), 85
RIBA Journal, 19, 25
RIBA Library Committee, 29
Rittel, Horst, 134, 212n6
RK algorithm, 178
Roe, Peter, 162, 164
Rowe, Colin, 29
Ruhnau, Werner, 51

Safdie, Moshe, 48, 74
San Francisco Poverty Program, 117
Schein, Ionel, 74
School Mathematics Project, 32–37, 40
Schulze-Fielitz, Eckhard, 48, 51, 67
Science Research Council, 166, 178
Scruton, Roger, 75–76
Second modernism, 7, 199n23
SEEK, 149
Semilattice, 117, 129
Sers, Philippe, 67
Set theory, 8, 27, 36, 39, 176
Shape grammars, 178–179, 183, 193, 218n175
"Sick Rose, The" (Blake), 122
Silverstein, Murray
Alexander and, 107, 116–118, 119f–120f, 124, 211n135
BART and, 116, 118
patterns and, 107, 116–124, 134, 211n135
Simmons, Kenneth, 117
"Skeleton of Science" (Boulding), 187
Skeletons
closets and, 190–191
computer graphics and, 189
computers and, 186–191
diagrams and, 186, 194
digital architecture and, 2, 6, 10, 185, 187
flesh and, 187–188
frameworks and, 53, 67, 74
ghosts and, 185–187
graphs and, 2–3, 6, 9–10, 185–193, 196
isomorphisms and, 188, 190
layouts and, 99, 103
modernism and, 187
networks and, 185–186, 190
pattern and, 187, 189, 191
systems and, 88–89, 186
SKETCHPAD, 154, 215n72
Smith, C. A. B., 57
Smithson, Alison, 29, 198n19
Smithson, Peter, 29, 198n19
Soft Architecture Machines (Negroponte), 181
Songs of Innocence and of Experience (Blake), 122
Soulis, George, 162
Southampton Mathematical Conference, 32
Space in the Home (Ministry of Housing), 46
Squaring the square, 46
SQUINT, 154, 182
Steadman, Philip
Architectural Morphology, 155–156
"The Automatic Generation of Minimum-Standard House Plans," 46
combinatorics and, 180
form and, 23f
The Geometry of Environment, 2f, 25–29, 36, 39, 43–48, 76, 98, 159, 161
"Graph Theoretic Representation of Architectural Arrangement," 156
isomorphism and, 75
layouts and, 98, 100f
possibility and, 155–161, 166
skeletons and, 188–189
Stiny, George, 178–179, 183
String diagrams
algorithms and, 82, 91, 93–94, 103, 129
layouts and, 82, 91–94, 98–99, 103, 129
Structuralism, 6–8, 24, 29, 39, 52, 122
Structure
abstract, 2, 5, 8, 128, 181
environmental, 116–124
hidden, 1, 15, 25, 30, 32, 74, 76
invariant, 4, 10, 173
mathematical, 113, 116, 118, 185, 190
new mathematics and, 29, 32, 36, 39
Studies in the Functions and Design of Hospitals (Nuffield Trust), 93
Sullivan, Louis, 172
Sutherland, Ivan, 154
Synge, John Millington, 19
Systematic Layout Planning (Muther), 93–94
Systems
aesthetics and, 202n47
combinatorics and, 180
layouts and, 97
patterns and, 106–110, 124
possibility and, 155–157, 161–164, 166, 178
skeletons and, 88–89, 186
of thought, 7, 9
"Systems Generating Systems" (Alexander), 124

Tabor, Philip, 97–101, 110–111
Tange, Kenzo, 48, 137
Tatton, W. E., 89
"Ten Year Program" (Alexander), 109
Tetris, 162
Textbooks
aesthetics and, 43, 48
algebra and, 24, 27, 35–36, 39
diagrams and, 27, 36, 43
The Geometry of Environment, 2f, 25–29, 36, 39, 43–48, 76, 98, 159, 161
graphs and, 25, 40, 43, 45f
graph theory and, 30, 36–37, 39
Land Use Built Form Studies (LUBFS) and, 27, 36, 46

Textbooks (*continued*)
March and, 25–29, 36–48
Martin and, 29
matrices and, 25, 27, 32, 35–36, 39, 47
networks and, 25, 36, 43
new mathematics and, 27, 29, 32, 35–39, 202n51
School Mathematics Project, 32–37, 40
set theory and, 27, 36, 39
Steadman and, 25–29, 36–48
symmetry and, 25, 27, 28f, 42–43
transformation and, 25, 27, 32, 35–40, 47
Thwaites, Bryan, 32, 36
Titan, 46–47
Topology
floor plans and, 157
Friedman and, 52
Hill and, 56
modern mathematics and, 8, 24
new mathematics and, 32, 35–36
transformation and, 41f, 161
Transformation
Chomsky and, 178
configurational engineering and, 161, 173
geometry, 25, 32, 35–36, 40, 161, 179
Klein, 35, 40
shape grammars and, 179
Stiny and, 179
Translation, 90, 128–130, 149, 202n47
Tree structures, 9, 117
Trinity Four, 46
Turner, John, 152
Tutte, Bill, 37, 46, 57, 162

United Nations, 134
University College London, 57
University of California, Berkeley
Alexander and, 107–117, 122–124
control and, 134
Montgomery and, 85
University of California, Los Angeles (UCLA), 157–158, 164, 166, 178
University of Cambridge
Alexander and, 14
Blanche Descartes and, 1
forms and, 19, 24
isomorphisms and, 75
LUBFS and, 15–16
textbooks and, 36, 46–47
Trinity College, 19, 46
University of Erlangen, 35
University of Manchester, 148
University of Michigan, 56–58
University of Waterloo, 161–164, 178
URBAN5, 149–150, 152, 189, 214n59
"Urban House, The" (project), 104
Urban mechanisms, 143, 143f–144f
Urban rule system, 110
Urban Systems Laboratory (USL), 214n71

Van der Ryn, Sim, 113, 116–117
Van King, Maren, 109, 112
VAX 11/780, 161
Vers une architecture (Le Corbusier), 48
Villes nouvelles, 146
Ville Spatiale, 51–53, 67, 74–75, 143, 146
Viollet-le-Duc, Eugène-Emmanuel, 42
Visualization, 4, 35, 48, 151–152, 198n11

W-algorithm, 125f–127f
Walkey, Ron, 124
Ward, Tony, 148
Webber, Melvin, 212n6
Weinzapfel, Guy, 150–152, 215n74
Welton Becket and Associates, 165
Whitehead, B.
Eldars and, 91–94, 98–103, 110–111, 156, 165, 208n47
layouts and, 90–94, 98–101, 103, 208n47
Whole Earth Catalog (Brand), 133, 136
Wittkower, Rudolf, 29
World War II, 7–8, 81, 186, 188
Wright, Frank Lloyd
Aline Devin House and, 42, 45f
"Eight Homes for Modern Living," 23–24
The Future of Architecture, 173
graphs and, 42–43, 45f
Imperial Hotel and, 48
Wurster, Bernardi & Emmons, 108–109

Xenakis, Iannis, 56

Yale University Computer Center, 157
Yantra, Mishra, 25, 26f
Your Own Native Architect (YONA), 138, 150–152, 181–182

The MIT Press would like to thank the anonymous peer reviewers who provided comments on drafts of this book. The generous work of academic experts is essential for establishing the authority and quality of our publications. We acknowledge with gratitude the contributions of these otherwise uncredited readers.

This book was set in Bembo Book MT Pro by the MIT Press. Printed and bound in Canada.

Library of Congress Cataloging-in-Publication Data
Names: Vardouli, Theodora, author.
Title: Graph vision : digital architecture's skeletons / Theodora Vardouli.
Description: Cambridge, Massachusetts : The MIT Press, [2024] | Includes bibliographical references and index.
Identifiers: LCCN 2023042997 (print) | LCCN 2023042998 (ebook) | ISBN 9780262049016 (hardcover) | ISBN 9780262379328 (epub) | ISBN 9780262379335 (pdf)
Subjects: LCSH: Architecture—Philosophy. | Graph theory.
Classification: LCC NA2500 .V37 2024 (print) | LCC NA2500 (ebook) | DDC 720.1—dc23/eng/20240103
LC record available at https://lccn.loc.gov/2023042997
LC ebook record available at https://lccn.loc.gov/2023042998

10 9 8 7 6 5 4 3 2 1